The Dialectical Tragedy of the Concept of Wholeness:
Ludwig von Bertalanffy's Biography Revisited

Exploring Unity through Diversity:
Volume 1

The Dialectical Tragedy of the Concept of Wholeness:
Ludwig von Bertalanffy's Biography Revisited
Volume 1: Exploring Unity through Diversity
Written by David Pouvreau
Translated from the French by Elisabeth Schober

Library of Congress Control Number: 2009923355

ISBN13: 978-0-9817032-8-2

© 2009 ISCE Publishing.

All rights reserved. No part of this publication may be reproduced, stored on a retrieval system, or transmitted, in any form or by any means, electronic, mechanical, photocopying, microfilming, recording or otherwise, without written permission from the publisher.

Printed in the United States of America

About the Book Series:
Exploring Unity through Diversity

Unity through Diversity is acknowledged to be the leitmotif of Ludwig von Bertalanffy's thinking. It is also the leitmotif of this series; that is, providing space for different perspectives while sharing a common goal in order to promote:

- Systems sciences, cybernetics and sciences of complexity as the most promising approaches towards global challenges humanity is facing in the new millennium;
- Transdisciplinarity and consilience throughout all scientific disciplines;
- The discussion and comparison of different schools of systems thinking;
- Attempts to unify systems thinking and to elaborate a metatheoretical framework;
- Systems history;
- Critical reflections of the development of systems thinking and the systems movement;
- Revisiting the goals of General System Theory as set by Ludwig von Bertalanffy, Anatol Rapoport, Kenneth Boulding and others;
- Social-scientific, that is, socio-economic, political, cultural and historical applications of systems thinking, including ecological and science-and-technology studies applications;
- Systems philosophy, and;
- Monographs or volumes of collected contributions in systems thinking as well as reprints of seminal works.

About this volume

The recent discovery of an archive full of personal documents of the philosopher and biologist Ludwig von Bertalanffy, founder of the General System Theory, paved the way for a reconsideration of important elements concerning his life and thought. This updated biography of a thinker, who is equally often cited as misjudged, takes into consideration all of his publications, his correspondence, as well as the secondary sources devoted to him, and attempts to reveal his richness and complexity to a general reader. This biography thus aims at initiating and promoting a study that is both critical and appreciative of his *oeuvre*. It equally seeks to navigate between two all too common pitfalls found in connection with von Bertalanffy: hagiographic temptation and reductive judgements, which are often ideologically motivated.

Wolfgang Hofkirchner
February, 2009

Bertalanffy Center for the Study of Systems Science (Vienna) Book Series
Exploring Unity through Diversity

Series Editor
Wolfgang Hofkirchner (Austria)

Editorial Board

Gabriele Bammer (Australia)
Yaneer Bar-Yam (US)
Gerhard Chroust (Austria)
John Collier (South Africa)
Yagmur Denizhan (Turkey)
Irina Dobronravova (Ukraine)
Klaus Fuchs-Kittowski (Germany)
Ramsés Fuenmayor (Venezuela)
Amanda Gregory (UK)
Ernesto Grün (Argentina)
Jifa Gu (China)
Debora Hammond (US)
Enrique G. Herrscher (Argentina)
Francis Heylighen (Belgium)
Cliff Hooker (Australia)
Magdalena Kalaidjieva (Bulgaria)
Helena Knyazeva (Russia)
George Lasker (Canada)
Allenna Leonard (Canada)
Gary Metcalf (US)
Gerald Midgley (New Zealand)
Gianfranco Minati (Italy)
Edgar Morin (France)
Matjaz Mulej (Slovenia)
Yoshiteru Nakamori (Japan)
Andreas Pickel (Canada)
Michel St. Germain (Canada)
Markus Schwaninger (Switzerland)
Len Troncale (US)
Martha Vahl (UK)
Gertrudis van de Vijver (Belgium)
Jennifer Wilby (UK)
Rainer E. Zimmermann (Germany)

The Dialectical Tragedy of the Concept of Wholeness:
Ludwig von Bertalanffy's Biography Revisited

Written by
David Pouvreau

Translated from the French by
Elisabeth Schober

**Exploring Unity through Diversity:
Volume 1**

Contents

Preface 1
Introduction and Acknowledgements 5

CHAPTER 1
Bertalanffy's Social Origins, Childhood and Adolescence 15

CHAPTER 2
His Years as a Student (1920-1926)

2.1 Studying at the University of Innsbruck
    (1920-1924) 19
2.2 Studying at the University of Vienna
    (1924-1926) 20
    2.2.1 The Curriculum Followed at the University of
    Vienna 21
    2.2.2 Relations with the Vivarium in Vienna 22
    2.2.3 The Doctoral Thesis 24

CHAPTER 3
Towards the Habilitation: The "Organismic" Theory (1926-1932)

3.1 Bertalanffy's First Writings after his Thesis
    (1926-1928) 27
3.2 The Reception of *Kritische Theorie der
    Formbildung* 29
    3.2.1 The Reception amongst Biologists 30
    3.2.2 The Reception amongst Philosophers,
        Psychologists and Psychiatrists 31
3.3 The Development of the "Organismic"
    Philosophy (1929-1932) 34
    3.3.1 Elaborating the "Organismic" Philosophy 34
    3.3.2 Biology as a Scientific and Cultural Crossroads
        and the Critique of Its Instrumentalizations 35
    3.3.3 Theoretical Biology 36

CHAPTER 4
The Reception of the "Organismic" Philosophy, the Habilitation and the First Years as Part-Time Lecturer in Vienna (1933-1937)

4.1 The Reception of the "Organismic" Philosophy 39
    4.1.1 The Reception in the Scientific World 39
    4.1.2 The Reception amongst Philosophers 40
    4.2 The Habilitation and a First Post as a Part-Time Lecturer 41
4.3 Bertalanffy's Situation and Works between 1933 and 1937 42
    4.3.1 The Theory of Organic Growth, a Concrete Expression of the Relevance and Fertility of "Organismic" Biology 43
    4.3.2 The Advent of Bertalanffy's New Connection to Mathematics 44
    4.3.3 An Increasingly Problematic Connection to the Neo-Positivists 46
    4.3.4 The Maturation of the "Organismic" Biology 47
4.4 An Opportunity to Go to the United States 49

CHAPTER 5
A First Trip to the United States (1937-1938)

5.1 Seven Months in Chicago: 51
The First Presentation of the General Systemology 51
    5.1.1 Some Contributions of Rashevsky's School 51
    5.1.2 A First Direct Contact with the "Integrative" Tradition of Chicago 52
    5.1.3 The First Presentation of a General Systemology .. 53
5.2 Bertalanffy's Reaction to the Anschluss and the Rest of His Stay in the United States 55

CHAPTER 6
Bertalanffy as a Biologist of the Third Reich (1938-1945)

6.1 Bertalanffy's Membership of the NSDAP (1938): His Reasons and His Benefits **59**
    6.1.1 A Return to Vienna in a Favorable Conjunction of Circumstances 59
    6.1.2 The Membership in the NSDAP 60
    6.1.3 The Complexity of Bertalanffy's Relation to National Socialism 61
6.2 The Promotion to the Title of "Associate Professor" (1938-1940) **66**
    6.2.1 A First Step: The Promotion to the Position of a Lecturer with Civil-Servant Status (1939) 67
    6.2.2 Bertalanffy as an Associate Professor: Some First Moves 68
    6.2.3 Bertalanffy's Work between 1938 and 1940: An Outline of a "Theory of Open Systems" and First Impacts of the Political Context 69
6.3 Bertalanffy as an Academic of the Third Reich (1940-1944) **71**
    6.3.1 The General Theory of Organic Growth, Beginnings of a "Dynamic Morphology" 71
    6.3.2 Bertalanffy's Explicit National-Socialist Commitment in his Writings in 1941 72
    6.3.3 The Art of "Manipulation" in order to Achieve One's Ends 74
    6.3.4 Bertalanffy's Scientific Activities between 1942 and 1944 76
    6.3.5 The Impact of Bertalanffy's Work during the War 78
    6.3.6 A Privileged Situation 79
6.4 The Family Disaster at the End of the War **80**

CHAPTER 7
His Last Years in Vienna: The "Denazification" and Its Consequences (1945-1948)

7.1 The Period Immediately after the War: Hope of Continuity 81
7.2 The Procedure of "Denazification" 82
    7.2.1 A Failed Attempt at Escaping the Procedure 82
    7.2.2 The Report of the Committee 84
    7.2.3 Purgatory and the First Desire to Emigrate 85
7.3 Bertalanffy's Work during a Period of Uncertainty 86
    7.3.1 His Actual Publications, Pending and in Progress 86
    7.3.2 The Alpbach Meeting: The Rebirth of the Project of a General Systemology and the Beginnings of a Philosophical Anthropology 87
7.4 The Results of the Procedure of "Denazification" 90

CHAPTER 8
Preliminaries of a Definitive Emigration (1948-1949)

8.1 An Invitation to London 93
8.2 An Intermediate Journey to Switzerland and a Scholarship for Canada. *Das Biologische Weltbild* and its Reception 94
    8.2.1 Bertalanffy's Activities in Switzerland 94
    8.2.2 "The Biological Conception of the World" 96
    8.2.3 The Reception of *Das Biologische Weltbild* 97
8.3 The Journey to Great Britain 99
    8.3.1 Another Ministerial Rejection of the Request for Promotion at the University of Vienna and the Renewal of the Leave of Absence at this University 99
    8.3.2 The Residence Permit in Canada 100

8.3.3 Bertalanffy's Activities in Great Britain: The Publicity of the Project of General Systemology in the Anglo-Saxon World and Its Reception 101

CHAPTER 9
Arrival in Canada: Montreal and Ottawa (1949-1954)

9.1 Six Months in Montreal 103
9.2 Bertalanffy's Preliminary Negotiations for His Transfer to Ottawa 104
9.3 Bertalanffy's Activities in Ottawa and the Reception of his Work 105
    9.3.1 The Development of the Theory of Open Systems 106
    9.3.2 The Beginnings of Bertalanffy's Entry into the Fields of American Psychology and Psychiatry 109
    9.3.3 The Pursuit of Experimental Research on Organic Growth 111
    9.3.4 The Beginnings of Bertalanffy's Research on the Pathology of Cancer 113
    9.3.5 Development and Promotion of the General Systemology 114
    9.3.6 The Aborted Project of the Philosophical History of Biology and the English Version of *Das Biologische Weltbild* *117*
    9.3.7 A Series of Conferences in the United States: Meeting Aldous Huxley 119
    9.4 Bertalanffy's Malaise in Ottawa and, More Generally, in America: An Aristocrat among Philistines 121
9.5 Bertalanffy's Conflicts with the University of Ottawa (1) 124
9.6 An Invitation to Stanford (California): The Prospect of an Institutionalization of the General Systemology 127
    9.6.1 Bertalanffy's Affinities with the Committee on the Behavioral Sciences of Chicago 127

- 9.6.2 Making Contact with Kenneth E. Boulding and Joining Forces in View of an Institutionalization of the general systemology 129
- 9.6.3 Bertalanffy within the Context of the Foundation of the Center for Advanced Study in the Behavioral Sciences (CASBS) 130

9.7 Bertalanffy's Conflicts with the University of Ottawa (2) 132
- 9.7.1 The Beginning of an Open Crisis: Bertalanffy Overtaken in the Run for Head of the Department of Biology 132
- 9.7.2 The Opposition of the University of Ottawa to Bertalanffy's Departure for Stanford 133
- 9.7.3 A Welcome Unlawful Dismissal and Its Advantageous Transformation 134

CHAPTER 10
The First Emigration to the United States and the Society for General Systems Research (1954-1955) at an Embryonic Stage

10.1 Bertalanffy's Theoretical Activities in Stanford 137
- 10.1.1 The Seminars of the CASBS in General and that of "General Systems" in Particular 137
- 10.1.2 The Systematized "Perspectivist" Philosophy .. 139
- 10.1.3 The Critique of the Social Role of the "Behavioral Sciences" and a New Moment for the "Dialectical Tragedy": The Individuality of Man as an "Ultimate Precept" 141

10.2 Bertalanffy's Organizational Work: The Formation of the Society for the Advancement of General Systems Theory and the Future Organization of the CASBS 144
- 10.2.1 A Scientific Society Dedicated to the Development of the General Systemology 144
- 10.2.2 Bertalanffy's Contribution to the Reflection on the Organization of the CASBS 146

CHAPTER 11
Three Years in Los Angeles and Two Trips to Europe (1955-1958)

11.1 Working next to Franz Alexander 149
   11.1.1 Bertalanffy's Living and Working
       Conditions in Los Angeles 149
   11.1.2 Bertalanffy's Commitment to Psychiatry:
       Studies on the Relations between
       Hallucinatory Drugs and Phenomena of
       Psychosis.. 151
   11.1.3 Some Developments of "Organismic"
       Psychology: Bertalanffy's Connection to
       Jean Piaget.. 152
   11.1.4 Successful Research on Cancer Diagnosis 154

**11.2 A First Trip to Europe and Lost Opportunities
to Return There for Good 155**
   11.2.1 The Prospect (without Follow-up) of a
       Professorship in Hamburg and the First
       Return to Europe.. 155
   11.2.2 The Two Lost Opportunities to Obtain a
       Professorship in Munich and Berlin 157

**11.3 The Foundation of the Society for General
Systems Research 158**

**11.4 The Conflicts with Alexander: A Second
Voyage and the Search for a Position in Europe. 160**
   11.4.1 The Strained Relationship between
       Bertalanffy and Alexander 160
   11.4.2 Return to Europe: Fruitless Negotiations
       in View of a Chair in Gießen................................... 161

**11.5 A Profound Weariness of America and
Repeated Efforts to Return to Europe 162**
   11.5.1 Bertalanffy Disgusted by the American
       Relationship to Science.. 163
   11.5.2 New and Unsuccessful Attempts to Return
       to Germany or Austria—The Need to Leave
       Los Angeles and Menninger's Rescue.................. 164

CHAPTER 12
Two Years in Kansas and a Third Voyage to Europe (1958-1960)

12.1 Bertalanffy's Activities at the Menninger Foundation 167
12.2 Bertalanffy's Difficulties at the Menninger Foundation 170
12.3 The Repeated Search for a Position in Europe ... 171

CHAPTER 13
The Second Canadian Period: Edmonton and Frequent Trips to Europe (1961-1969)

13.1 Bertalanffy's Liberty and Deployment in Edmonton 175
    13.1.1 Bertalanffy's Administrative Activities between 1961 and 1963 176
    13.1.2 A Period of Synthesis and Intellectual Liberty .. 177
13.2 "Organismic" Psychology: General Systemology as the Matrix of a "Humanistic" Science 181
13.3 Regular Trips to Europe 184
13.4 The Interest Bertalanffy Aroused in the Eastern European Countries 186
13.5 His Last Years in Edmonton 187

CHAPTER 14
A Last Return to the United States: Twilight Years in Buffalo (1969-1972)

14.1 Teaching at SUNY 191
14.2 The Critique of the Mathematic Tendencies of the General Systemology 192
14.3 The Disastrous Year of 1971 193
14.4 Steps towards a Nomination for the Nobel Prize and Bertalanffy's Last Days 194

Epilogue 197

APPENDIX A: The Complete Works of Ludwig von Bertalanffy 199
Posthumous Publications 222

APPENDIX B: Secondary Sources on Ludwig von Bertalanffy's Life and Thought 225

Index 235

About the Author

David Pouvreau concluded his studies in mathematics with the agrégation, a high-level competitive examination for recruitment of teachers, and holds an advanced studies degree (D.E.A.) in history of sciences from the School of Advanced Studies in Social Sciences (EHESS) in Paris, carried out under the supervision of Jean Dhombres and in association with the Alexandre Koyré Centre. Under the same conditions he is currently working on his doctoral thesis, which deals with the history of Ludwig von Bertalanffy's General System Theory. He has already published several papers on the latter subject, in the USA and in France.

About the Bertalanffy Center for the Study of Systems Science (BCSSS)

Given the global challenges of today, systems science is needed more than ever. Yet system theory is not mainstream. The objective of the BCSSS is to inspire the development of systems science. The BCSSS aims at the advancement of scientific research in the field of systems thinking. In particular, it revisits General System Theory (GST) as founded by Ludwig von Bertalanffy and others in order to reassess it in the light of today's global challenges and to illuminate the course of development systems science has taken since. The BCSSS is open to cooperation with every person or organisation supporting the same aim. The BCSSS owns the Ludwig von Bertalanffy archive and possesses a collection of publications of the systems movement.

Website: http://www.bertalanffy.org/

For further information please contact:

> Hofkirchner Wolfgang, Prof. Dr.
> ICT&S Center
> University of Salzburg
> Sigmund-Haffner-Gasse 18
> 5020 Salzburg
> Austria
>
> Phone: +43 662 8044 4821
> Fax: +43 662 6389 4800
> Wolfgang.Hofkirchner@sbg.ac.at

Preface

Ludwig von Bertalanffy seems not to be forgotten. References are still made to him. As Michel Saint-Germain, Professor at the Faculty of Education, University of Ottawa, writes in the preface for the French edition of this book in 2006[1], according to the database *Web of sciences*, 642 publications throughout the last five years have thus made reference to Bertalanffy at least in their bibliography; in the database *Scopus* one finds 1350 documents of the same type; there are 532 in the *Science Citation Index (expanded)*; and one finds ten in the *Philosopher's Index*. This might be an indication for classifying Ludwig von Bertalanffy, in contemporary thinking, and remembering him, in the first place as biologist and founding father of General System Theory.

However, Ludwig von Bertalanffy was not a mere naturalist. From the outset Ludwig von Bertalanffy was interested in biology just as well as in philosophy—in science just as well as in humanities. He was born into the post-*fin-de-siècle* and post-World-War-I intellectual atmosphere of Vienna which, finally, was the medium to help give birth to the cross-disciplinary endeavours and weaving-together of systemic, evolutionary, cybernetic and informatic trains of thought.

Bertalanffy was a critic of the current development of civilization and, by that, a critic of the currents of science. In the late sixties he even became critical of the then *en-vogue* trend in education to institutionalize system theory and cybernetics in the form of new spe-

---

[1]. The French edition was published electronically on the website of the BCSSS: Pouvreau, D. (2006). *Une biographie non officielle de Ludwig von Bertalanffy (1901-1972)*, ISBN 9783200008403.

cialties, thereby producing new kind of nerds instead of broad minds.

His historical importance is thus twofold: Ludwig von Bertalanffy was a global thinker concerned with the fate of humanity and he was a thinker in complexity, the latter entwined with the first.

It was not without his personal predilection for cutting across disciplines and not without the revolutionary incubation milieu of Vienna of that time that enabled him to overcome the mechanicism/vitalism divide in biology and to carve out the cornerstones of General System Theory for another period of human evolution to come. It is these features of his thinking that make it worth revisiting when investigating the foundations he laid for subsequent systems thinking and the anticipations he achieved of the most recent developments in the sciences of complexity and systems philosophy as well as insights not yet resumed for one or another reason[2].

How to conceive of wholeness might be considered the centre around which Ludwig von Bertalanffy's thinking revolved. Bertalanffy's emphasis on the hierarchical nature of the structure processes form in living systems coincided with the Nazi ideology of the *Führer* principle. Bertalanffy did not refrain from seizing the opportunity to propagate his scientific idea by paying lip service to its dictatorial derivate. This should become his major concession to Nazism with regard to his theoretical work. But as we can easily see, it is not inherent to his theory.

David Pouvreau gives a definite account of the changing historical, socio-political, and scientific con-

---

[2]. See, e.g., my reflections on the occasion of the the First World Congress of the International Federation for Systems Research, Kobe, Japan, in Hofkirchner, W. (2005). "Ludwig von Bertalanffy—Forerunner of Evolutionary Systems Theory", in J. Gu and G. Chroust (ed.), *The New Role of Systems Sciences For a Knowledge-based Society*, ISBN 4-903092-02-X.

texts as well as of the personal circumstances of Ludwig von Bertalanffy's life from which his ideas originated one after the other.

This present English edition is different from the French one in that it has been updated regarding the main text as well as the appendixes. It was quite a collective effort to finalize this edition. Not only did Elisabeth Schober an excellent job when going beyond the mere task of translation but also Manfred Drack and Birgit Zehetmayer helped correct imprecision, in particular, where familiarity with the Austrian environment is needed.

Wolfgang Hofkirchner
Vienna, October 2008

Introduction and Acknowledgements

For three quarters of a century one of the remarkable aspects of scientific thought and its relations to philosophy and ideology has been the recurrent return, in different forms, of what one can qualify as the systemic paradigm. More exactly, the return of a will to establish such a paradigm, which is supposedly alternative to a "mechanistic" vision of the world and revolutionizing not only science as such, in its cognitive contents and epistemological foundations, but also its social, political and ethical dimensions.

The justification or the rejection of this will, as well as the reasons for such a recurrence, constitute in my opinion a major challenge in the history of the sciences and, to a larger extent, in the history of ideas. To grasp it in its complexity is, in my view, not possible without a thorough examination of the *oeuvre* and the life of a philosopher and biologist who constitutes in all respects a central figure of systems science and systems philosophy: Ludwig von Bertalanffy (1901-1972). He was in fact a major shaper and a founding father of the systems movement, which originated in the 1950s in the United States. But perhaps he also represents the agent of this movement who expresses best the diversity of its dimensions and its aporias.

Insofar as Bertalanffy is regularly cited in systems literature as an indispensable reference and as he is also regularly attacked on the part of the critics of this literature, one could assume that this person is well known. However, this is not at all the case. On the one hand his thought is often only known superficially, by the former as by the latter—two major reasons for this are the diversity and multiplicity of the dimension of his writings and the fact that a large proportion of them is written in German, a language that is mastered less and less frequently. On the other hand, the knowledge of his life, important for the comprehension of his ideas, has been not only insufficient and imprecise until now, but almost always distorted by an "official" biography.

The primary biographical source concerning Bertalanffy is established by himself, through individual subtle remarks in his writings. A first supplement was contributed by his wife in 1973 (M. Bertalanffy, 1973). And it was in close collaboration with her that the American science journalist Mark Davidson wrote and published a biography of Bertalanffy in 1983, which has represented since then the major reference in this matter. However, this "biography" is not really a biography at all: it constitutes much rather a hagiography that falsifies history when it compromises too much the glorious image of the hero, whose path it pretends to trace. In Davidson's defence one has to acknowledge that Bertalanffy and his wife Maria themselves have neatly taken care of providing the elements of this "official" history already in the post-war period. In this respect also Gerhard Nierhaus needs to be mentioned, who, in a long article published in 1981 in Germany, had already procured numerous elements of this hagiography, right after having contacted Maria von Bertalanffy (Nierhaus, 1981).

Veronika Hofer, in her doctoral thesis defended in Vienna in 1996, was the first to uncover the main part of Bertalanffy's life that had been kept hidden, by revealing his compromising relations with National Socialism. She also obtained interesting unpublished documents about his relations with the famous "Vienna Circle". In these two cases she was brought to examine a small but important part of his correspondence, which was available in certain Viennese archives—here again, she was the first to have done that. Unfortunately, she limited herself basically to these aspects, strangely ignoring certain fundamental elements of Bertalanffy's first works (although she was the first who took them into consideration), along with the period after his emigration to America. And she based herself on them in order to argue for an interpretation of Bertalanffy's thought that is, in my view, simplistic and adjusted and that too often disregards its complexity and its dialectical aspect.

The doctoral thesis defended in Münster by Sabine Brauckmann in 1997 also contributes several new, even though much less substantial, biographical elements. Brauckmann elicits them from several letters written by Bertalanffy, which she was able to procure from the United States. Surprisingly though, she does not seem to know Hofer's work: she does not use any of Hofer's data and analyses. And she takes over to a very large extent the "official" history.

The last work that examines Bertalanffy's biography (in one of its parts) is the doctoral thesis of Debora Hammond, revised and published in 2003. Here again, this work is based almost entirely upon Davidson's hagiography. She certainly also brings in several new elements, taken from letters exchanged between Bertalanffy and certain other initiators of the systems movement. But the fact that she does not include Hofer's work and numerous works of Bertalanffy that were published before 1945 lead her (just like Brauckmann before) to imprecisions and to at times questionable interpretations of his *oeuvre*; and, to a larger extent, to a negligence of certain dimensions of the history of systems theory regarding the object of its study—its social implications.

The main difficulty concerning Bertalanffy's biography is due to the complete destruction of his personal archives in 1945 in the disaster of the end of the Second World War. Until 2004 this difficulty was increased by the fact that one did not know where the personal archives of the later period were situated either.

A major event in this respect, which overturned the facts of this problem, took place in March 2004 in Buffalo. On the occasion of a clearance sale of a second-hand bookstore, the American historian of photography and daguerrotypist Rob McElroy found six curious banana boxes in the middle of about 20,000 books piled up in this shop (Chroust & Hofkirchner, 2006). And he discovered that these heavy and bulky boxes contained the archive in question: it turned out later that he was dealing with no less than 579 letters exchanged between Bertalanffy and various scientists, philosophers and institutions between 1946 and 1972; some remains of his library: 212 books, conference proceedings and reviews; several dozens of reprints and photocopies of articles; various handwritten notes; and the majority of his articles published throughout his whole career.[1] We still do not know who sold his archive to this bookshop, but it seems that it had been there, without ever being discovered, for the duration of at least ten or even twenty years—maybe even already since shortly after the death of Maria, on 6 October 1981. The reason for this is that the banana boxes, *a priori* judged little propitious for a quick sale, were passed over and, finally, forgotten in a corner of the bookshop, where they quickly became hardly accessible any more.[2]

McElroy, who understood the importance of his discovery, communicated it already in April 2004 to the American poet Tom Mandel, whom he knew to be associated with the systems movement. On the initiative of the professors Elohim Jiménez-López, Gerhard Chroust (University of Linz and *International Federation for Systems Research*—IFSR), Gerd B. Müller (*Konrad Lorenz Institute for Evolution and Cognition Research* and University of Vienna) and above all Wolfgang Hofkirchner (Vienna University of Technology), who directed them from the Austrian point of view, negotiations with Elroy were then undertaken in the course of the year 2004 in

---

[1]. More exactly the boxes contained about 90% of his articles published after 1945, and about a third of those published before that. Few of his books, however, are present in this archive.

[2]. Personal report by R. McElroy.

order to transfer this archive to Vienna. Particularly on account of the financial support of the private foundation of a nephew of Ludwig von Bertalanffy these negotiations were successful, and the archive arrived in Vienna on July 6.[3] On this occasion Hofkirchner founded the *Bertalanffy Center for the Study of Systems Science* (BCSSS) on December 17, 2004, in order to promote the examination of this archive and, more generally, the study of the historical and epistemological foundations of systems science. The archive is accommodated in the Department of Theoretical Biology at the University of Vienna, which, despite a number of organizational changes in the Faculty of Life Sciences, constitutes a continuity with the Institute of Zoology at which Bertalanffy had been working until the end of the Second World War. Bertalanffy's archive has been entirely categorized and classified there since March 2006.

Through Michel Saint-Germain, professor at the University of Ottawa and first author of a doctoral thesis on Bertalanffy in 1979, I was able to get in contact with Wolfgang Hofkirchner by the end of the year 2005. Five weeks of work at the BCSSS, between February and May 2006, allowed me to examine the entire archive in detail. Furthermore, Saint-Germain also allowed me to have access to an administrative file concerning Bertalanffy at the University of Ottawa, where he had worked during a crucial period of his career. This file contains thirty letters, which I have also analyzed in detail. The biography which I present here is largely the fruit of all these works. But it also rests upon a more general doctoral study about the history of Bertalanffy's thought, which I carry out within the framework of the *École des Hautes Études en Sciences Sociales* (EHESS, Paris).

The present work has the objective of synthesizing in the first place all the biographical data provided

---

3. Chroust and Hofkirchner, 2006. Personal report by E. Jiménez-López.

by the previous studies, of correcting their errors and complementing them substantially. These studies contradict themselves sometimes, and Bertalanffy's archive allowed me to shed light on the corresponding points. The three main contributions of my biography concern on the one hand the decade that has remained mostly obscure until now, which reaches from the end of the war to Bertalanffy's departure to the United States at the end of 1954; on the other hand, his largely "non-official" ideas that concern the "critique of culture" and the scientific world; and finally, the logical, psychological, intellectual, and institutional ideas that explain the development of his career. This biography also adds several important supplements to Hofer's study as regards the relations between Bertalanffy and National Socialism.

Even though my original concern consisted in providing all the previously available and assured data concerning Bertalanffy's life, it nevertheless appeared impossible to me to present these elements separate from his works. I have also tried to integrate a presentation of his major ideas in my biographical reconstruction, by respecting as scrupulously as possible the chronology of their emergence. I have equally mentioned the most important scholars and philosophers who played a role in the genesis of his thought. This biography has thus the complementary aim of introducing a reading of his *oeuvre* in accordance with his development, something that is unfortunately only seldom the case in other studies dedicated to him and which renders them refutable to a certain number of commentaries. This integration of Bertalanffy's writings in my biography rests inevitably on an interpretation determined by their genesis. I insist, however, on the fact that I have kept the interpretative work to a minimum. I avoided as much as possible the critical analyses that, although necessary, are not part of my goals. My work was driven by a constant concern for "objectivity" and by the wish to provide the data prior to any correct interpretation of Bertalanffy's work, in its details as in its entirety. I generally leave the effort of appropriating these data in a critical manner to the interested reader. The only point for which I deemed it indispensable to violate this discipline a little concerns Bertalanffy's relations with National Socialism. This is in fact a very delicate matter, upon which hasty conclusions can be drawn easily. I thus wanted to introduce to this subject some elements of reflection that allow us to have a nuanced and objective judgment. By choosing the title of my biography, taken from an expression used by Bertalanffy himself in his correspondence, I wished to provide a frame of analysis of his entire thought that seemed relevant to me in this regard—by limiting myself to emphasizing the aspects that justify this frame of analysis, without undertaking a systematic analysis.

This biography accordingly is meant to be above all a basis for future work, which can and should be improved. For this reason I have provided the first comprehensive bibliography of Bertalanffy's works, which is regularly referred to in the body of the text.[4] I also considered it useful to attach a bibliography of works already dedicated to Bertalanffy—this list, however, is not exhaustive.[5] It is indeed remarkable that these works hardly ever take into consideration those that preceded them. One can add two more to this list, still in progress and both in the field of history of sciences: that of Roberto Manzocco, University of Pisa, which focuses on the so-called "new vision of man" developed by Bertalanffy and which is in particular meant to provide important elements concerning the network of relations after the war;

---

4. In 2000 Sabine Brauckmann provided a bibliography that, although it constituted a remarkable progress in relation to that included in the collection edited by W. Gray and N.D. Rizzo in 1973, was still incomplete.

5. I did not list the dozens of "reviews" of Bertalanffy's books, since they were only very short. I have identified all the works that, to my knowledge, discuss Bertalanffy's ideas in a substantial manner.

and mine, which deals with the origins, the genesis, the updating, and the heritage of his *Allgemeine Systemlehre* ("General Systemology").[6]

I would certainly not have been able to successfully complete this biography without the initial help of Michel Saint-Germain, whom I should like to thank in the first place. We were just about to make inquiries about Gisèle von Bertalanffy (who lives in Canada, Winnipeg) in order to find out where the archive of her father-in-law might be located, when he got the idea, seeing that our search would certainly be vain, to contact Wolfgang Hofkirchner—whom he had met in Vienna in 2001 to celebrate the centenary of Bertalanffy's birth. We were thus able to find out quickly about the famous discovery of Rob McElroy, whom I would also like to thank for having described to me in detail the conditions of this discovery.

I also want to express my thanks to all the members of the BCSSS, in particular to Wolfgang Hofkirchner and Gerd B. Müller, for the warm and stimulating welcome they gave me during the time I spent in Vienna, as well as for the excellent working conditions they have offered me there. In this regard I also thank Heidemarie

---

6. The term "general system theory", chosen by Bertalanffy, was and remains in itself a source of confusion, both in the usage of the word "theory" as in the singular used for the word "system". In fact, this term did not satisfy Bertalanffy himself either and was chosen for lack of a better one so as to translate the original German expression "*Allgemeine Systemlehre*". With Manfred Drack (University of Vienna), I prefer translating it by "general systemology", which seems to better account for the fact that it is neither a scientific theory nor a philosophical doctrine: it integrates these aspects as particular moments of a more general construction that has also logical and methodological dimensions. This expression is also historically justified: it seems legitimate to hold the "*Systematologie*" of Johann H. Lambert (1787) for the first attempt that prefigured, as much in its intentions as in its contents, Bertalanffy's "systemology".

Pollack, secretary of the *Department of Theoretical Biology*, who has facilitated considerably the practical conditions for this work.

I also thank the mathematician and historian of exact sciences Jean Dhombres, my supervisor at the EHESS, for his patient rereading and his judicious advice.

Finally, I would like to thank the French Ministry of Research for having awarded me, thanks to the support of Jean Dhombres and Wolfgang Hofkirchner, a grant that allowed me to carry out my research in Vienna under optimal circumstances.

CHAPTER 1
Bertalanffy's Social Origins, Childhood and Adolescence

The Bertalanffy family is descended from Hungarian nobility. Its oldest known ancestor was Lieutenant Isaac Bertalanffy, rewarded for his military exploits by Rudolph I, King of Hungary in the sixteenth century. The second family member who distinguished himself in Hungarian history was the Jesuit Paul Bertalanffy (1706-1763), who wrote the first book in the Hungarian language—a prayer book. The Austrian branch of the family was founded by Karl-Josef von Bertalanffy (1833-1912), the mark of nobility *von* being then added to the name as a sign of belonging to the aristocracy at the end of a request acceded to by imperial decision. Although originally a lawyer and in spite of the conventions related to his noble rank, Karl-Josef directed an itinerant troupe of actors and was later appointed director of the theatre of Klagenfurt at its inception. He is still known as an audacious innovator in the history of Austrian theatre. His eldest son Gustav (1861-1919) was administrator of an Austro-Hungarian railway company; he also became an advisor to the emperor at the end of his career. In 1895 he married Charlotte Vogel, the daughter of an imperial advisor who also owned a print shop in Vienna—she was then 17 years old. Karl Ludwig von Bertalanffy was born of this union on September 19, 1901, in Atzgersdorf, a village near Vienna (De Vajay, 1973; Nierhaus, 1981: 144. Brauckmann, 1997: 1; Hofer, 1996: 6).

Being a single child, Karl Ludwig was pampered by his mother. His parents got divorced in 1911, and each of them remarried shortly afterwards. He lived with his mother and her second husband—Eduard Kaplan, also an administrator of a railway company—in a liberal household open to a wide circle of friends coming from the artistic, literary and scientific world.

Until 1911 he was educated by private tutors. After that he entered the Karl-Ludwig Gymnasium in Vienna, where he was the classmate of the future composer Hanns Jelinek (1901-1969).[1] Parallel to his schooling, he immersed himself in the classical literature of antiquity, reading—in the German tradition—Homer, Plato, Virgil and Ovid in the original. He also wrote poems, a play about Cesare Borgia and a novella entitled *Der neue Tristan* ("The New Tristan")—these writings disappeared in 1945. Karl Ludwig was very early on interested in biology: he possessed his own laboratory with microscope, where he practiced vivisection and plant and animal anatomy. This interest was stimulated by the visits to the controversial (neo-Lamarckian and socialist) biologist Paul Kammerer (1880-1926), then neighbour and friend of the family. And very early he studied the theoretical texts of Lamarck, Darwin and Marx (Brauckmann, 1997: 1; Hofer, 1996: 7; M. von Bertalanffy, 1973: 32-33; Davidson, 1983: 49).

Karl Ludwig's father died in 1918. For their part, for fear of a communist revolution in Austria, his mother and stepfather sold all their land and quickly found themselves ruined by post-war hyperinflation. The family moved to Zell am See in late 1918. From then on Karl Ludwig was no longer enrolled at school. During the next two years he pursued his studies in an autodidactic manner, and then took his *Matura* (school-leaving exam) as an individual candidate and passed it with distinction (Hofer, 1996: 8) in July 1920. This was also the period when his mother, in memory of a love affair, began to call

---

1. Correspondence between L. von Bertalanffy and H. Jelinek (21/03/1966-19/11/1966), BCSSS archive.

him only Ludwig.[2] And quickly the first name Karl disappeared from familiar and public usage.

---

[2]. Letter from L. von Bertalanffy to H. Jelinek (24/04/1966), BCSSS archive.

CHAPTER 2
His Years as a Student (1920-1926)

2.1 Studying at the University of Innsbruck (1920-1924)

Bertalanffy began his university studies in autumn 1920, at the Faculty of Philosophy of the University of Innsbruck. He studied there until the summer of 1924, following a curriculum that was standard in German-speaking countries but which can surprise people coming from other traditions. The livelihood of his family and the costs his studies seemed mainly secured by the guesthouse his mother had in Kufstein (Tyrol).

In the first year he took his courses in botany, philosophy and art history. In botany he attended the courses in experimental morphology of Emil Heinricher (1856-1934), and those in plant physiology of Adolf Sperlich and Adolf Wagner. The latter, a defender of a form of vitalism, was particularly interested in natural philosophy and theory of knowledge, his ambition being to obtain the foundations for biology that would provide him with an epistemological unity. Through its wholly experimental approach that unifies morphology and physiology, just as through its opening onto philosophical questions, this school of botany exerted a significant influence on Bertalanffy. As to his philosophical training, it took place with Franz Hillebrand (1863-1926) in logics, theory of knowledge and psychology; and with Alfred Kastil (1874-1950) in philosophy of religion and metaphysics—both philosophers coming from the school of Franz Brentano (1838-1917). In art history finally, Bertalanffy attended courses in ancient art with Heinrich Sitte (1879-1951), in medieval art and graphic arts with Moritz Deger (1868-1939), and in philosophy of art history with Heinrich Hammer (1873-1953).

The next three years, interrupted during the second semester of the year 1923 for unknown reasons, were exclusively devoted to philosophy and art history. His professors were still Hillebrand and Kastil in philosophy; and Sitte, Dreger and Hammer in art history. In the second semester of the year 1924 a weekly course in history of pedagogy (Hofer, 1996: 8-10) may be added.

Equally during this period Bertalanffy immersed himself in a study of "German mysticism" (from Master Eckhart, Nicholas of Cusa, Paracelsus and Jakob Böhme to Johann Wolfgang von Goethe and Arthur Schopenhauer), that gave him the material for his first article in 1923. Amongst others he had a passion for Oswald Spengler (1880-1936), whose *The Decline of the West*, which conceives of cultures as organisms that are moved by immanent laws of evolution and can be the subject of a "comparative morphology", left an indelible mark on his mind. Moreover, in December 1924, he published a long "introduction" to this book.

2.2 Studying at the University of Vienna (1924-1926)

In April 1924 Bertalanffy met Maria Bauer, who had come from Vienna to recover from the Spanish flu in the guesthouse of Bertalanffy's mother. Their discussions focussed immediately on his reflections on mysticism, philosophy of history and history of art. From then on they were inseparable. She would recount later that she must have "fallen in love with her brain"[1]. When Maria had to go back to Vienna in September 1924, Ludwig accompanied her. They lived with Maria's father, and

---

1. Davidson, 1983: 51-52, quoting Maria von Bertalanffy.

Chapter 2: His Years as Student

Ludwig now enrolled at the University of Vienna. How he managed to subsist and to finance his studies remains unknown—probably at the same time owing to his mother's help and his accommodation with Maria.

2.2.1 The Curriculum Followed at the University of Vienna

Bertalanffy studied only two semesters at the University of Vienna.

In the first semester of 1924/1925 he showed no more than three preceding years of particular interest for biology: he enrolled in four courses in history of art and in German literature.

In the second semester he did not enrol. The main reason for this was that he married Maria on March 1, 1925. She abandoned her own studies (mathematics and medicine) and became from then on not only his wife, but also his collaborator, his chauffeur, his secretary, and his "manager", since Ludwig, following Klemenz W.F. von Metternich (1773-1859), a diplomat and conservative Austrian chancellor, thought that creative individuals must devote themselves exclusively to the things they can do rather than to waste their time with tasks others can accomplish. Furthermore, one pressing question concerning his orientation arose: should he enter on a career as a philosopher or as a biologist? His wife advised him to choose the path of biology, because there was a better chance of getting work in this field and because "a biologist can use what he knows to be a philosopher, but it cannot work the other way round" (Davidson, 1983: 52-54; Hofer, 1996: 10-12).

But Bertalanffy did not yet wholly take this advice to heart and had trouble deciding. In the first semester of 1925/1926 he enrolled exclusively in philosophy. Four hours a week he attended the course "Philosophy of the Renaissance and Rationalism" of the neo-Kantian professor Robert Reininger (1869-1955), who had been at the head of the Philosophical Society of Vienna since 1922. Bertalanffy equally attended a seminar given by the latter on reading philosophical texts. For another five hours a week he went to the lectures of Moritz Schlick (1882-1936) on "Logic and Epistemology" and took part in a seminar organized by this neo-positivist philosopher and founder of the famous "Vienna Circle" in 1929. He also attended the course in "Epistemology of Historical Sciences" of Viktor Kraft (1880-1975)—another future member of the Vienna Circle. And he attended two courses closely related to the philosophy of biology: one on "The Philosophy of Henri Bergson" (by a man named Garbei) and the other on "Myth as a Biological Creation" (by a man named Schneider; Hofer, 1996: 12). Nevertheless, he showed little interest in these different teachings and went there often only to sign the list, sometimes even letting his wife do it in his place (M. von Bertalanffy, 1973: 33).

2.2.2 Relations with the Vivarium in Vienna

During this period Bertalanffy also established relations, directly or indirectly, with members of the *Vivarium* in Vienna, an "experimental institute of biology" [biologische Versuchsanstalt] founded in 1903 through private funds by the zoologist Hans Przibram (1874-1944) and the botanists Leopold von Portheim (1869-1947) and Wilhelm Figdor (1866-1938). This institute was oriented towards interdisciplinary and experimental studies of morphogenetic problems that abandon descriptive and comparative biology in favour of a search for causal explanations. In particular, these scientists were carrying out experiments on regeneration, transplantation, the influence of temperature on biological processes, and experiments in relation to comparative embryology (Hofer, 2000: 144-145). The institute as a whole was marked by an opposition to Darwinian and neo-Darwinian theses. It certainly accepted evolution, but neither the idea that the principle of natural selection was sufficient to explain it, nor the preformationist concept of heredity defended by August Weismann (1834-1914): its dominant notion is that the organism needs to be understood as a system that establishes an active relation with its environment, and that the morphogenesis needs to be seen as the product of epigenetic processes (Hofer, 2000: 146-147).

After Kammerer, a member of the institute since 1914, had resigned in the wake of a scandal caused by the allegation of the falsification of his experiments that were aimed at demonstrating the inheritance of acquired characteristics, Paul A. Weiss (1898-1989) succeeded him in 1924. In his doctoral thesis of 1922, published only in 1925, Weiss attacks the theory of tropism of Jacques Loeb (1859-1924), which reduces animal behavior to an expression of reflexes conditioned by the physico-chemical environment of the organism. He responds to it with a systemic theory that shows that this behavior, principally faculties of self-regulation and adaptation, have to be understood in view of the dynamics inherent in the organism and can be explained by the principles of optimality. But he does not restrict his theory to animal behavior: he extends its scope by outlining a general theory of organized systems. His ideas owe much to "Gestalt theory", developed in Germany by the Psychological Institute in Berlin, where Wolfgang Köhler (1887-1967) since 1919 strived to extend its scope beyond psychology in the direction of a systemology—a term which he coined in 1927—by establishing general systemic principles that could be applied to psychology as well as to physics and biology. Weiss's theory quickly became the frame of reference of the Institute of Zoology at the University of Vienna, and Bertalanffy soon showed interest in it, as well as in the works of Köhler. He made contact with Weiss himself. From now on the two young researchers regularly had heated discussions, at least until 1926, concerning the concept of the system and its ap-

2.2.3 The Doctoral Thesis

Between late 1925 and February 1926 Bertalanffy merged his interests in metaphysics, the theory of knowledge, philosophy of culture, and biology, when working under the direction of Schlick and Reininger on his doctoral thesis in philosophy, entitled *Fechner und das Problem der Integrationen höherer Ordnung—Ein Versuch zur induktiven Metaphysik* ("Fechner and the Problem of Integration of Higher Order—An Attempt at an Inductive Metaphysics"). Furthermore, he published a first article on the metaphysics of life ("Die Einheit des Bildungstriebes", "The Unity of the Drive of Education") as of late 1925.

By retracing its history in a critical manner, his thesis deals with the controversial problem of knowing to what extent, on a metaphysical as well as on a scientific level, the idea that "supra-individual entities" that are composed of living organisms can be understood as "integrations of higher order" is justified, since they have their "individuality" and their own laws; and more generally, to what extent the world as a whole can be considered a hierarchy of levels of organization. By drawing on a study on Gustav T. Fechner (1801-1887), Rudolf H. Lotze (1817-1881), Eduard von Hartmann (1842-1906), and Henri Bergson (1859-1941), but also on recent work on experimental biology, Bertalanffy attempts to justify, while reformulating them, Fechner's basic ideas. He shows that the questions at issue can receive a positive and fruitful answer within a scientific framework, provided that general principles of organization are substituted by analogies taken from psychology. And he suggests that the development of inductive metaphysics is possible on this scientific basis.

Chapter 2: His Years as Student

Bertalanffy's thesis was approved on February 21, 1926. The defence took place on March 1. While Reininger judged the accomplished work as "excellent", Schlick limited himself to "sufficient +", as he believed that it lacked coherence, clarity and logical force, while recognizing a considerable knowledge in both biological and philosophical questions (Hofer, 1996: 12-13). At the end of his life Bertalanffy himself considered this thesis as "one of the worst he had encountered in a long academic life" (M. von Bertalanffy, 1973: 33). The fact remains that it foreshadowed the topics and the outline of his systemic thought.

A momentous event concerning his family life also took place while he was writing his thesis: his wife gave birth to their only child Felix on January 20.

CHAPTER 3
Towards the Habilitation: The "Organismic" Theory (1926-1932)

Even though promoted to the rank of doctor of philosophy, Bertalanffy had for all that not secured his future. He had yet to obtain his habilitation (postdoctoral qualification), which was only possible subsequent to substantial and recognized publications. Such publications were moreover a means to increase his livelihood, which, although other sources are not exactly known, was certainly under control.

3.1 Bertalanffy's First Writings after his Thesis (1926-1928)

In 1926 Bertalanffy published two articles concerning the philosophy of art history, one on the philosophy of nature of the poet Friedrich Hölderlin (1770-1843) and another on the renewal of mysticism in Russia. His only article concerning biology was a first entry into the debate between "mechanists" and "vitalists", then raging in the philosophy of biology. He concluded it by affirming the necessity of passing beyond the controversy by taking a middle course.

In 1927 he dedicated himself more fully to the philosophy of biology. Except for a text on "Classical Utopia", which criticizes the contemporary classicistic dreams of spiritual renewal and which opposes to them, following Spengler, the thesis of an ineluctable decline of the West, he in fact published five articles dominated by contemporary debates in biology this year. By leaning on the epistemological upheavals in contemporary physics in particular, he argues there in favour of a "methodological vitalism" that is opposed both to reductionist approaches to the organism and to a metaphysical vitalism considered scientifically fruitless.

This direction continued in 1928 with two new articles. One of them deals with von Hartmann: Bertalanffy underlines there at the same time the point of his inductive metaphysics and the relevance of his criticism of Darwinism, of which he would afterwards keep appropriating arguments by adapting them in a modern form. But 1928 was more generally a symptomatic year of the diversity of Bertalanffy's interests since 1920.

He published in fact also in the same year an article in the field of philosophy of art history, commenting favorably on the *oeuvre* of Josef Strzygowski (who had been one of the readers of his thesis), whose project was to get art history outside of a purely descriptive orientation in order to confer to it the quality of an "exact" science.

As for his interest in mysticism, it led to the publication of his first book, still in 1928, exclusively devoted to Cardinal Nicholas of Cusa (1401-1464), in which he translated those parts of Cusa's works that he considered highly interesting and commented on them. All his life he remained deeply committed to the latter, whom he considered the father of the modern spirit. He praised his synthesis of a mystical intuition of the organic unity of the world and of the affirmation of a power of reason that culminates in mathematics; and he appropriated his doctrine of "learned ignorance" [*De Docta Ignorantia*], his vision of a reality that, as a "coincidence of opposites" [*coincidentia oppositorum*], is both imperceptible in its foundation and partially accessible in a "perspectivist" form.

1928 was finally the year when Bertalanffy's first work dedicated to biology was published: *Kritische Theorie der Formbildung* (English title: *Modern Theories of De-*

velopment). In this work he outlines various, past or current, theories of morphogenesis, which he submits to a critical analysis whose aim is on the one hand to show the need for a theoretical approach to biological phenomena, and on the other hand to reveal an "organismic" philosophy of living things as the only tenable one. He advances and develops with this philosophy the idea that organization is a property of living systems, of which there is no equivalent amongst non-living beings; that it is accessible through a scientific study because it can be explained on the basis of "forces" that are immanent to these systems; but that it cannot be understood through analytical processes that, by definition, disregard this, and that it thus needs to resort to specific, "integrative" [ganzheitlich], biological categories. This implies an at least provisional independence of biology in relation to the physico-chemical sciences, with regard to its concepts, principles and laws. *Kritische Theorie der Formbildung*, however, is not limited to epistemological considerations. The radical criticism of the dominance of an "atomist", determinist and reductionist—in short, "mechanistic"—approach to the organism constituted for Bertalanffy only a necessary moment, certainly essential, of a general reorientation of a vision of the world that affects not only biology and science as a whole, but also culture: for Bertalanffy, who believed that all science was necessarily a particular incarnation of a certain vision of the world, criticism of "mechanism" was simultaneously and necessarily also one of an "atomized" and "mechanized" society.

3.2 The Reception of *Kritische Theorie der Formbildung*

This book, because of its logical clarity, its mastery of the problem of morphogenesis, the theoretical course that it outlines in order to lead biology out of the controversies that tear it apart, and the philosophical perspectives it opens, was highly successful and rapidly made Bertalanffy known in Germany and England, amongst philosophers as well as biologists.

3.2.1 The Reception amongst Biologists

Its publication by Julius Schaxel (1887-1943), professor at Jena, marks Bertalanffy's official entry into the movement of those biologists who sought to develop a theoretical approach that opened a middle course between "mechanism" and "vitalism", which was considered untenable and fruitless respectively. Schaxel, who popularized the term *organismisch* ("organismic") in 1919, coined by the entomologist Ludwig Rhumbler (1864-1939) in 1906, published in fact all those who pursued the goal he had set himself. Bertalanffy heavily fed on the authors in question—besides Schaxel himself, these were, amongst others, Weiss, Kurt Lewin (1890-1947), Emil Ungerer (1888-?), Karl Sapper (1866-1945), Johannes Reinke (1849-1931), Alexander Gurwitsch (1874-1954), and Friedrich Alverdes (1889-1952). Moreover, it was Schaxel who had invited Bertalanffy to write his *Kritische Theorie der Formbildung*, after having read his articles on philosophy of biology of the two preceding years with great interest.

Bertalanffy was of course very favorably received in Germany and in Austria by the biologists inclined to reject "mechanism" and to adopt an "integrative" approach. In particular, he won the support, at the University of Vienna, of a former student of the sociologist Max Weber (1864-1920), the renowned morphologist Jan Versluys (1873-1939), who was then at the head of the department of zoology; and also that of the great Viennese botanist Richard von Wettstein (1863-1951).

Bertalanffy also attracted the attention of the members of the *British Theoretical Biology Group*, which included notably the biochemist Joseph Needham (1900-1995) and the embryologists Edward S. Russell (1887-

1954), Joseph H. Woodger (1894-1981), and Conrad H. Waddington (1905-1975). Having worked one year at the *Vivarium* in Vienna in 1926 and taking a great interest in the developments in logistics and its possible application in the theory of biology, Woodger was not only familiar with the experimental and integrative approach developed by the Viennese zoologists, but also with the Viennese neo-positivist movement, whose principal actors he had frequented and remained in contact with. He introduced Bertalanffy to his colleagues and soon began to translate his book into English. Conversely, Bertalanffy familiarized himself via Needham and Woodger with the English emergentist and organicist movement, especially with the theories of C. Lloyd Morgan (1852-1936) and Alfred N. Whitehead (1861-1947).

3.2.2 The Reception amongst Philosophers, Psychologists and Psychiatrists

On the part of the philosophers, Bertalanffy first of all attracted the interest of the neo-positivist circles in Vienna and above all in Berlin, by participating in some of their seminars, giving talks on several occasions in 1928, 1929 and 1931 (Hofer, 1996: 235-240), and publishing by far the longest article of the first volume of their journal *Erkenntnis* in 1931. His contacts with certain members of these circles even seem to have been friendly,[1] and it is certain that Hans Reichenbach (1891-1953), leader of the Berlin circle,[2] was among them. He fully shared their interest for the critical analysis of scientific concepts; their concern for the unity of science; and also their will to defend its value in a context of cultural crisis, where

---

1. F. von Bertalanffy, "Vorwort," in L. von Bertalanffy, *Das biologische Weltbild*, Wien, Köln, 1990 (2nd edition), p. viii, quoted in Hofer, 1996.
2. Letter from H. Reichenbach to L. von Bertalanffy (24/02/1951), BCSSS archive.

neo-romantic, existentialist, mystic, and nihilist tendencies united so as to call it into question. Bertalanffy nevertheless remained fundamentally outside of these circles. He objected to their radical empiricism; their rejection of metaphysics, which he regarded as sectarian and naïve; and their myth of a refined science from which all relations to ethical and religious values are banned.

In the late 1920s, he also attended and sometimes spoke at the seminar of the psychologist Karl Bühler (1879-1963), whose criticism of the most radical tendencies of the Vienna Circle, embodied by Rudolf Carnap (1891-1970) and especially by Otto Neurath (1882-1945), prefigured to a certain extent the conflict that in the 1950s and 1960s was to oppose the advocates of a "humanist" psychology (among others, Bertalanffy) to those of a "behaviourist" psychology. Bertalanffy became in fact a lasting friend of Karl Bühler and especially of his wife Charlotte (1879-1974), who was also a psychologist and an inspiration to "humanist" psychology in post-war America.[3]

He introduced himself simultaneously to *Gestalt* theorists, whose influence on his thought he acknowledged: Köhler (under whom he intended at one point to write his postdoctoral thesis) and Lewin frequented in fact Reichenbach's circle in Berlin; Lewin even gave a seminar on Bertalanffy's book (Hofer, 2000: 153-154). The latter also introduced himself to the renowned psychiatrists Kurt Goldstein (1878-1965) and Alexander Herzberg (1887-1944), who also attended meetings of the circle and defended holistic conceptions in their field that converged clearly with his.

More discreet but more profound than with the neo-positivists were Bertalanffy's connections with different neo-Kantian circles, whose attention he also soon attracted. He spoke in seminars of the *Kant-Gesellschaft*

---

[3]. Correspondence between L. von Bertalanffy and C. and K. Bühler, BCSSS archive.

("Kant Society") and published in the important journal *Kant-Studien* ("Kant Studies"). Through Reininger, who integrated the "organismic" philosophy of biology of his student in his own reflections, Bertalanffy got in contact with Hans Vaihinger (1852-1933), who welcomed him several times at his home in Halle and with whom he maintained friendly relations (M. von Bertalanffy, 1973: 34). The "fictionalist" philosophy of this important figure in German neo-Kantianism had a lasting influence on Bertalanffy, who dedicated two articles to him, one in 1929 and another in 1932, and who at the end of his life still considered him as his "paternal friend" (L. von Bertalanffy, 1966: 115). Bertalanffy also familiarized himself at this time, without meeting them, with the thought of other great neo-Kantian philosophers or philosophers of neo-Kantian origin, such as Heinrich Rickert (1863-1936) or Nikolai Hartmann (1882-1950)—whose theory of the stratification of the real and of the "fundamental categories" he considered a forerunner of his general systemology on a metaphysical level—and probably Ernst Cassirer (1874-1945)—with whom the affinities and influences were soon to be profound and mutual.

Still in the field of philosophy, Bertalanffy also got in contact with the philosopher of science Bernhard Bavink (1879-1947), with whom he corresponded and of whom he appropriated some major ideas. He equally became friends with Richard Wahle (1857-1932) in Vienna, exposing himself to a philosophy that revived the sceptical tradition (M. von Bertalanffy, 1973: 34).

3.3 The Development of the "Organismic" Philosophy (1929-1932)
3.3.1 Elaborating the "Organismic" Philosophy

The period from 1929 to 1931, which thus saw Bertalanffy progressively establishing himself in certain academic circles, despite the fact that he still did not have any position at a university, was dedicated to the elaboration of his "organismic" philosophy. It was marked primarily by a broadening of the biological problems considered: not only embryology, but also the physiology of the metabolism and the problem of evolution became the subjects of his critical reflections. It was also marked, undoubtedly because of his exposure to neo-positivism and neo-Kantianism, by a development as far as the theory of biological knowledge was concerned, which led Bertalanffy in particular to clarify his conception of the relations between science and metaphysics. Finally, it was devoted to the search for general principles of biological organization, intended to show the potential of the "organismic" philosophy, its capacity to generate an actual theory of the organism—i.e., for Bertalanffy himself, a hypothetical deductive system where biological laws derive logically from hypothetical principles and are empirically testable.

During this prolific period he published twenty articles, almost all of them devoted to the philosophy of biology. Until 1930 the very modest income he derived from these publications, as well as from *Kritische Theorie der Formbildung*, permitted him more or less to secure the livelihood for himself and his family.

3.3.2 Biology as a Scientific and Cultural Crossroads and the Critique of Its Instrumentalizations

In 1930 he published his second book devoted to biology: *Lebenswissenschaft und Bildung* ("Life Science and Education"). More explicitly than in his previous writings, Bertalanffy insists there on the central role of biology in contemporary science and, more generally, in contemporary culture. Not only because the categories of "wholeness" or "organization", *par excellence* those of biology, seemed then (in the German-speaking world) paradigmatic in medicine, psychology, "cultural sciences" [Kulturwissenschaften] and even, to a certain extent, in physics, but also because political and educational conceptions claimed to be based on biology. Bertalanffy analyses in this book the dangers of such a situation, related in his opinion primarily to the lack of theoretical foundations in biology. He mainly attacks the ideological instrumentalizations of biology, whether in a so-called Lamarckian justification of socialism or in an alleged legitimization of the ethics of competition, of eugenics or of racism on the basis of Darwinism and the theory of heredity. Something he did less out of ethical concern—he was not opposed to eugenics or to racism in general, only to a radical eugenics that is scientifically unfounded—than out of the will to save biology from a mixture of genera that could only be harmful to it and to highlight the detrimental consequences that one could expect from ideological points of view, where rationality is only an illusion. For Bertalanffy, life science could only healthily assume the role of a "crossroads" and vector of a reorientation of the vision of the world by rooting itself in a theoretical biology worthy of the name. And the purpose of an "organismic" philosophy was to give him the means of satisfying this demand.

3.3.3 Theoretical Biology

Between 1930 and 1932 Bertalanffy benefited from a scholarship awarded by the Austrian section of the *Notgemeinschaft der deutschen Wissenschaft* ("Association for the Support of German Science"), thanks to its president von Wettstein. This scholarship gave him more financial comfort and allowed him to work in an untroubled manner on the first volume of his *Theoretische Biologie* ("Theoretical Biology"), published in late 1932.

In this treatise Bertalanffy systematically outlines and justifies his "organismic" conception as a means to pass beyond the controversy between "mechanism" and "vitalism" and as a matrix of a truly theoretical biology in analogy to theoretical physics—that is to say, a nomothetic science of organic nature. He identifies organization as the fundamental problem of biology and shows that this problem cannot be dealt with within the framework of an "analytical" ("atomistic") epistemology that claims to comprehend the object of biology only from the knowledge of its parts considered in isolation. "Organismic" biology is therefore the appropriate perspective, the "positive" research programme and the method that this problem requires; its purpose is the elaboration of a "systemic theory" of the organism that provides the laws of biological systems in the form of "statistics of higher order" [Statistiken höherer Ordnung] by disregarding the complexity of the underlying processes.

An original contribution of this book is the formulation of two major "principles", whose validity is not limited to the living *organism*, but extended to all *organized* (biological) *systems*, from the cell to biocoenoses; their purpose is to play a role for biology analogous to that of the principles of movement for Newtonian physics. Any biological system constitutes first of all an "open system" that remains in or moves towards a "dynamical pseudo-equilibrium", away from the thermodynamic equilibrium (a state of "rest" which signifies its death):

Chapter 3: Towards the Habilitation

this system is the seat of a specific order that can only be created and can only continue to exist through a flux of matter and energy; and its "equilibrium" is in fact maintaining a permanent non-equilibrium. In addition, the order in question constitutes a "hierarchical order" of parts and processes that needs to be understood in two ways. First of all in a "static" sense: a biological system results from the integration of a multiplicity of levels of organization; from the atom to life communities [Lebensgemeinschaften], any entity of a particular level is subject to the laws that govern the superior level and presents characteristics with regard to the entities of the inferior level that cannot be deduced from the characteristics of the latter when considered in isolation. And secondly in a "dynamic" sense: this integration, designed on the model of epigenetic development, is based on the dynamic interaction within the global system; it is progressive and implies a closer and closer dependence of each part on the logics of wholeness. Any organized system is initially an "equipotential" wholeness, undifferentiated and in this way capable of an optimal regulation; it is the seat of a process of "segregation" and "differentiation" of its parts during which their functions are specialized; but this "disintegrative" tendency is interdependent with an opposing process of "progressive centralization" during which a chain of subordination is generated, which leads to the existence of "dominant parts" and, ultimately, to an existence of a sub-system that controls the whole.

In considering all the problems of contemporary biology, with the notable exception of genetics, Bertalanffy tried hard to show that these principles were relevant on all levels of organization and could help to unify the various biological fields in the same theoretical perspective. He supported the idea that the autonomy of biology in relation to physics was justified as long as its framework, obviously too narrow in order to account for such principles, was sufficiently extended; and he suggested furthermore that theoretical biology was likely, in fact, to lead to an extension of physical concepts so that the problem of the "reduction of biology to physics" was ultimately a purely rhetorical problem: it depended on the extension of meaning one attaches to the expression "physical science".

CHAPTER 4
The Reception of the "Organismic" Philosophy, the Habilitation and the First Years as Part-Time Lecturer in Vienna (1933-1937)

4.1 The Reception of the "Organismic" Philosophy
4.1.1 The Reception in the Scientific World

With his *Theoretische Biologie*, Bertalanffy established himself as one of the major philosophers of biology of his time on an international level. But the reception of his book was mixed among German and Austrian biologists and doctors, although unanimity as to his remarkable mastery of all biological problems and his erudition prevailed.

He attracted first of all the criticism of biologists and philosophers of biology of "vitalist" inspiration, such as Alois Wenzl, who considered his "organismic" perspective as a sophisticated variant of "mechanism" similarly to *Gestalt* theory. And he had to suffer the inverse criticism on the part of eminent biologists such as Max Hartmann (1876-1962), who saw in his thought a pernicious form of "vitalism".

Nevertheless, he also received important support. For example, from the renowned neurophysiologist Albert Bethe, who wrote a laudatory review of his book in 1933; from the botanist Fritz Gessner (1905-1972), who equally published a favourable review of his book in 1934 and subsequently became his best friend until the end of his life;[1] or also from the physician Heinrich Zimmermann, who saw in his conceptions a liberating course for medicine. In addition, Woodger completed his translation of *Kritische Theorie der Formbildung* and inserted in it that of the first, epistemological, part of *Theoretische Biologie*. Both were published in 1933 under the title *Modern Theories of Development*. These last two books immediately received laudatory reviews from members of the *British Theoretical Biology Group*, in particular in the prestigious journal *Nature* (by Needham). Bertalanffy was therefore soon known not only in Great Britain, but also across the Atlantic. He attracted the attention of the physico-mathematician Nicolas Rashevsky (1899-1972), who also aimed at the development of a theoretical biology worthy of the name—in the form of a "mathematical biophysics"; but also that of pure biologists, who even integrated *Modern Theories of Development* in their courses.[2]

As regards the physicist Pascual Jordan (1902-1980), one of the fathers of quantum mechanics, he was also strongly interested in Bertalanffy and based himself explicitly on him from 1934 onwards in order to justify the holistic theory of the relations between quantum physics and biological organization, which he tried to develop since 1932—while conversely his work stimulated Bertalanffy's reflections.

4.1.2 The Reception amongst Philosophers

In the field of philosophy, the reception of *Theoretische Biology* was on the whole excellent, which was not surprising in an ideological-political context where "philosophies of wholeness" [Ganzheitsphilosophien] played

---

[1]. Correspondence between L. von Bertalanffy and F. Gessner, BCSSS archive.
[2]. Letter from R.O. Earl to L. von Bertalanffy (05/01/1950), BCSSS archive.

an important role in the legitimization of National Socialism. Bavink wrote a very favourable review of the book and even advocated the creation of a professorship of theoretical biology for Bertalanffy. As for Carl Fries (1895-1982), he saw in Bertalanffy the incarnation of a renaissance of natural philosophy and devoted a book to him that appeared in 1936, where he celebrated the "new path created through the alliance between science and philosophy"[3] that is outlined in *Theoretische Biologie*. More significant still was the considerable interest that Bertalanffy aroused in Cassirer. The author of *Die Philosophie der symbolischen Formen* (*The Philosophy of Symbolic Forms*), an imposing monument of twentieth-century philosophy, developed in fact in the last part of his work a philosophy of biology that relies heavily and explicitly on the perspective outlined by Bertalanffy in his *Theoretische Biologie*.

4.2 The Habilitation and a First Post as a Part-Time Lecturer

From then on Bertalanffy had the problem of his habilitation and that of finding a post, having just turned 31 years old. All the more so as he found himself in early 1933 without scholarship (as a result of political disagreements between Germany and Austria) and lived then only on the earnings of his latest book. On June 27, 1933, he submitted an application for his habilitation in "Theoretical and General Biology" [Theoretische und allgemeine Biologie] to the council of professors of the University of Vienna. His *Theoretische Biologie* was presented as a postdoctoral thesis (*Habilitation*). The commission, which notably comprised Schlick, Reininger and Versluys, unanimously approved this habilita-

---

3. Fries, 1936: 72: wie "aus einem Zusammengehen der Naturwissenschaft und Philosophie ein neuer Weg gefunden wird".

tion, but demanded that it should only read "theoretical biology". Bertalanffy was the first to be habilitated under this title in Vienna; the Faculty of Philosophy was convinced of the necessity of his approach.

Nevertheless, Bertalanffy had to wait still almost one year before he got a modest post as a *Privatdozent* (a lecturer without salary) in theoretical biology in the department of zoology at the University of Vienna. His habilitation to such a post was only issued on May 9, 1934 (Hofer, 1996: 14, 16; Brauckmann, 1997: 4; M. von Bertalanffy, 1973: 37).

4.3 Bertalanffy's Situation and Works between 1933 and 1937

Bertalanffy began his teaching at the University of Vienna in October 1934. Each of his courses was limited to two hours a week—the rest of his time was devoted to research. In the first semester of 1934/1935 he taught an "Introduction to Theoretical Biology", followed by a course on the "Facts and Theories of Morphogenesis" in the second semester. In the first semester of 1935/1936 he taught "A Short Overview of the Major Biological Theories"; and in the second semester "An Overview of the Basic Concepts of Physiology". His course in the first semester of 1936/1937 concerned the "Physiology of Phenomena of Excitement and of Behaviour"; and that of the second semester was entitled "Atomistics of Wholeness in Modern Biology".

Bertalanffy's publications between 1933 and 1937 were less numerous than in the preceding period, but they were marked by a significant qualitative development. There were only eight articles between 1933 and 1934, and then two, in Hungarian, between 1935 and 1936. This period ended, however, with the publication of a voluminous and dense treatise in 1937: *Das Gefüge des Lebens* ("The Structure of Life").

His financial difficulties were considerable during this period. Thanks to the intervention of the physicist Jordan in Germany and of his colleague von Wettstein in Vienna, he managed to obtain another scholarship of the *Notgemeinschaft der deutschen Wissenschaft*, but only in 1936 (Hofer, 1996: 16-17).

4.3.1 The Theory of Organic Growth, a Concrete Expression of the Relevance and Fertility of "Organismic" Biology

From 1933 onwards, Bertalanffy concretely tried to implement the "organismic" perspective and programme which he had outlined in 1932. From then on he in fact wanted to demonstrate its fertility. He therefore focussed on the problem that he considered the most suitable for that purpose: that of morphogenesis. Half of his articles in 1933 and 1934 were already devoted to it: they deal with morphology—in particular with the concept of homology—and with organic growth.

Bertalanffy actually concentrated quickly on the latter problem, which he considered central to that of morphogenesis. He first became attached to the study of the global growth of the organism, for which he elaborated a first mathematical theory in 1934, establishing its relationship with metabolic processes. And in a second phase, from 1935 onwards, he studied the problem of the relative growth of organs. For various reasons, the choice of this research topic was judicious from the perspective of his "organismic" biology. It allowed him first of all to implement his conception of the organism as an open system, the phenomenon of growth being seen as the result of an excessive assimilation of matter by the organism in dissimilation. It also gave him an opportunity to call attention to the benefit of understanding biological phenomena from the point of view of a "statistics of higher order", by showing that the knowledge of complex physico-chemical processes underlying the phenomenon under consideration is not necessary for producing overall laws about it. The study of this phenomenon allowed him also to achieve the synthesis of two biological fields whose unity had then not yet been built and which were traditionally even antagonistic: physiology (of the metabolism) and morphology.

But another reason for this interest in phenomena of growth is that they had the advantage of being quantifiable and were then part of the few biological phenomena that had been treated mathematically. His theory of global growth allowed Bertalanffy to show, in the spirit of his "organismic" philosophy, that "exact" systemic laws can be formulated and, in particular, to "prove" that such phenomena as the equifinality of growth (the independence of the final state in relation to the initial state and to the routes undertaken to reach it), regarded by some as major arguments in favour of vitalism, are in fact systemic characteristics that are immanent to organised matter and are mathematically predictable.

4.3.2 The Advent of Bertalanffy's New Connection to Mathematics

An advancement of Bertalanffy's connection to mathematics was perhaps the most striking novelty of his thought in this period. This development, initiated in fact around 1930, contrasted with his previous positions on this subject, which, in the tradition of Goethe and Bergson, had rejected the very idea of a mathematical biology. It resulted from a tension between this "neo-romantic" inspiration and an "organismic" programme that, from 1928 onwards, aimed at developing a nomothetic biology whose "exact" laws were deductible from general principles.

A major cause for this development was Bertalanffy's discovery of various recent works that introduced a mathematical processing of certain biological questions, in which he recognized not only a potential

Chapter 4: The Reception of the "Organismic" Philosophy...

for achieving his "organismic" programme, but also for expanding its scope outside of biology. During this period certain specific issues—such as cell division and nervous excitation—were the subject of mathematical approaches based on models that were to some extent "organismic", which thereby aroused Bertalanffy's attention. The physico-mathematical theory of cell division of Rashevsky, for example, was developed on the basis of the model of the open system. In addition, the recurrent validity of the same elementary mathematical "law" (the "allometric" equation) on the various levels (the phylogenetic, ontogenetic, biochemical and metabolic levels) of the problem of morphogenesis, posed the problem of its origin and appeared to him as the formal expression of the existence of general principles of organization. Numerous interactions between theories of global and dynamic organic growth of populations contributed also to this acknowledgement of the existence of "isomorphisms" and to the necessity of justifying its origin and relevance. Bertalanffy was particularly interested in the works of Alfred J. Lotka (1880-1949) and Vito Volterra (1860-1940), who presented mathematical laws for certain ideal biocoenotic systems that converged in many regards with his "organismic" perspective. He was very impressed and influenced by Lotka. The latter adopted the perspective of a "statistics of higher order" and understood the evolution of biological systems in general from the concept of the open system (without phrasing it like that); above all, he developed the principles of a "general kinetics" and outlined a "general theory of states" [Allgemeine Zustandslehre] that applies to all dynamic systems that exist in nature. Lotka seemed thus to Bertalanffy to provide a formal framework adapted to the implementation of a generalization of the "organismic" perspective and programme, to which he aspired in fact since he had formulated them, without having the scientific means of justifying them.

4.3.3 An Increasingly Problematic Connection to the Neo-Positivists

Bertalanffy's contacts with the neo-positivists and neo-Kantians, at least before 1933, were not, far from it, unrelated to this evolution of his connection to mathematics either, which essentially resulted from a maturation of his conceptions in the theory of knowledge.

His connection to the latter between 1933 and 1937 unfortunately remains totally unknown. As regards his relationship with the former during the same period, it remains obscure. He seems to have maintained friendly relations with the members of the circles of Vienna and Berlin, even organising meetings at his home with some of them.[4] His correspondence with Neurath also showed a great mutual respect and revealed a keen interest in Bertalanffy's ideas on the part of the defenders of the "scientific conception of the world", which found expression in repeated invitations to their conferences (Hofer, 1996: 242-243). Bertalanffy, who declined the invitation in 1936, contributed to the third conference in Paris in 1937. But at the same time, his ideas were strongly criticized by at least two members of the Vienna Circle: Philipp Franck (1884-1966) in 1932 and Felix Mainx (1900-1983) in 1935. And at the same time, Neurath and Carnap advocated a physicalist unity of science, adding thus to the already significant divergences between the neo-positivist views and those of Bertalanffy. He seems in fact to have had very contrasting connections to these circles, remaining always at their periphery, at least from the point of view of philosophical orientations, and unilaterally breaking off the relations during his stay in the United States in late 1937, since the divergences had

---

4. F. von Bertalanffy, "Vorwort," in L. von Bertalanffy, *Das biologische Weltbild*, Wien, Köln, 1990 (2nd edition), p. viii, quoted in Hofer, 1996. Letter from H. Reichenbach to L. von Bertalanffy (24/02/1951), BCSSS archive.

appeared too important to him. In addition, the assassination of Schlick by one of his students in 1936 marked the beginning of a dissolution of the Vienna Circle and of a general emigration of German and Austrian neo-positivists to less hostile climes (especially to the United States), which naturally resulted in a weakening of the already complicated ties with Bertalanffy.

4.3.4 The Maturation of the "Organismic" Biology

In two of his articles published in 1933 and 1934 Bertalanffy develops a critical approach to the problem of evolution, initiated in 1929 and continued in his *Theoretische Biologie*. He opposes to the Darwinist and neo-Darwinist theories of evolution the phenomenon which he would describe from 1949 onwards as "anamorphosis", the tendency towards an increase of order and complexity which manifests itself in ontogenesis as much as in phylogenesis; a tendency that, according to him, makes hazard and natural selection certainly necessary but not sufficient conditions for evolution. He suggests that evolution is co-determined by systemic laws that are immanent to living organisms, which explains in particular the phenomena of convergent evolution. This criticism is even intensified in *Das Gefüge des Lebens*, where Bertalanffy for the first time consistently takes into account the results of research in genetics. He strives to show there that, genetic research, far from being opposed to his "organismic" perspective, requires it. For him, phenomena of pleiotropy and polygenesis[5] indeed reveal that the genome constitutes in itself an organized system. And to explain the phenomenon of co-adaptation,[6] the crux

---

5. In pleiotropy one gene influences several traits, whereas in polygenesis several genes influence one trait.
6. "The term coadaptation refers to the mutual adaptation of [...] genes, so that a gene may be favored by selection if it is in the same individual as a particular gene at another locus," Marc Ridley (2003),

of selectionist and mutationist theories of evolution, he postulates (long before their actual discovery) the existence of patterning genes, i.e., genes that coordinate the action of groups of other genes. *Das Gefüge des Lebens* is in fact more generally the expression of a maturation of Bertalanffy's "organismic" thought.

This maturation manifests itself especially in the close relationship that is established between his two "organismic" principles, which function in two ways. On the one hand, any organized system is henceforth clearly understood as an open system whose opening is precisely the *condition* of the possibility of hierarchization, of "self-organization". On the other hand, any organic hierarchy, from the cell to biocoenosis, is itself understood as a hierarchy of open systems in "dynamical equilibrium", where what appears as a stable configuration at a given level, a "structure", is in fact maintained by a continuous change of its components and is designed as the expression of a "dynamical equilibrium" of components on the subordinated level of the hierarchy.

Another aspect of this maturation is the clear formulation of three fundamental moments of his "organismic" philosophy. The first one, the "integrative" conception of any organized system as a "unified whole", had already been explained in 1932. The second one also, but it was henceforth more elaborate. It is a *dynamic* conception of organized systems, which is opposed to a "static" conception that reduces them to an arrangement of fixed structures. The rigid forms are dissolved in a flux of processes, and the opposition between structure and function vanishes. The order generated by the dynamic interaction of parts is declared crucial. And the "organismic" principle of hierarchization is completed by the introduction of the concept of "progressive mechanization", which on the one hand refers to the specialization

---

"Co-adaptation," in *Evolution*, http://www.blackwellpublishing.com/ridley/a-z/Coadaptation.asp.

of parts, to the relative autonomy they acquire in the course of the process of differentiation, and to the subsequent diminishing of the capacities of self-regulation of the system; and on the other hand reveals "mechanized" structures of an organized system such as secondary products, derivatives, of the primary dynamic order. Finally, Bertalanffy explained in 1937 for the first time the third fundamental moment of his "organismic" philosophy: that of the "primary activity" [primäre Aktivität] of the organism. His insistence on this concept had indeed been inherent in his thought from the very beginning. It is linked to his constant efforts to develop an "organismic" alternative to the neo-Darwinian theory of evolution, and to his discovery of holistic approaches in medicine, neuro-psychiatry, neuro-physiology and behavioural sciences, which he considered as the most remarkable and promising developments that converged with his own ideas. To the "mechanistic" conception of the organism as a "puppet" that responds passively and univocally to the demands of its environment, embodied in the behaviourist "stimulus-response" model, he opposed the conception of the organism as a system that has an autonomous activity, which "metabolizes" all the actions of its environment in accordance with its own logic.

4.4 An Opportunity to Go to the United States

While he approached the age of 36 and was still only a *Privatdozent*, Bertalanffy considered in 1937 that this status did not conform to the richness and importance of his works. It is true that they were publicly acknowledged by biologists (Alverdes, Gessner, Needham, Woodger, etc.) as well as by physicians (Zimmermann, Karl E. Rothschuh, etc.), philosophers (Bavink, Fries, etc.), and at least one renowned physicist (Jordan). He therefore decided to address a request to the dean of the Faculty of Philosophy of the University of Vienna, in which he asked for his promotion to the rank of "associate professor" [*außerplanmäßiger Professor*].

But even though the file was taking its course and had not yet been settled, he seized an opportunity: on the recommendation of Rashevsky he obtained a scholarship from the Rockefeller foundation in order to work one year in the United States from October 1, 1937, to September 30, 1938. From the Austrian point of view he was "to study American developments in biology with a view to the application at Austrian universities" (M. von Bertalanffy, 1973: 43); and from the American point of view, to work on mathematical biology and the department of biophysics created in 1935 by Rashevsky at the University of Chicago. The scholarship was granted to him because his work was in line with the interest of the mathematician Warren Weaver (1894-1978), who had important functions in the Rockefeller foundation and advocated the development of mathematical biology.

Bertalanffy thus took one year of leave of absence from the University of Vienna and left for the United States with his wife, leaving his son in the care of his grandparents. His choice had a downside: his request for promotion to the rank of "associate professor" received a positive response on 3 March 1938, nine days before the annexation of Austria by Hitler… But his nomination was suspended a few days later because of his absence (Brauckmann, 1997: 5, 8).

CHAPTER 5
A First Trip to the United States (1937-1938)

5.1 Seven Months in Chicago: The First Presentation of the General Systemology

Bertalanffy's journey to the United States began with a long stay in Chicago, with the exception of a few short trips, which lasted until May 4, 1938.

5.1.1 Some Contributions of Rashevsky's School

Bertalanffy and his wife lived close to the university and enjoyed the hospitality of the small circle of Rashevsky's assistants, including Herbert D. Landahl (1913-), Alvin M. Weinberg (1915-), Alston S. Householder (1904-1993) and John M. Reiner.

Bertalanffy's relationship to Rashevsky as such was rather strained. The two researchers had some theoretical differences: Rashevsky had of course used the model of the open system; he also represented one of the pioneers of mathematical biology and of a theoretical approach to biology; but at that time he had adopted an openly reductionist approach to biological problems, which was diametrically opposed to Bertalanffy's „organismic" conceptions. However, this was only a minor reason. The tension between them was mainly due to a conflict between two very strong personalities that manifested itself in several altercations (Brauckmann, 1997: 7).

This conflict did not prevent Bertalanffy from having conversations with Rashevsky's assistants, which were decisive for the development of his thought. They concerned the theory of open systems and the unsuitability of classical thermodynamics for understanding biological systems. From his discussions with Reiner, he derived the possibility of formalizing the evolution of open systems in a very general way with the help of equations of partial derivatives also known as ‚general transport equations', and that of discussing *a priori* the characteristics of their solutions—an approach which caught up with Lotka's *a priori* discussion about certain characteristics of dynamic systems based on the formalism of differential systems. From his discussions with Householder, Bertalanffy derived also, more specifically, the possibility of establishing a formal link between the open character of a system and its equifinality: the latter seemed thus no longer a necessary characteristic of single living systems, but a general systemic characteristic.

Even before these discussions, which were initiated at the beginning of his stay, had led to a formal theory of open systems, they functioned, in conjunction with Bertalanffy's entire writings and readings of the last four years, as a catalyst for the "theory" that was to establish his reputation: his general systemology.

5.1.2 A First Direct Contact with the "Integrative" Tradition of Chicago

Beyond the circle around Rashevsky, which was only one incarnation amongst others, Bertalanffy discovered an "integrative" tradition at the University of Chicago, of which he had already heard something in Vienna through biological literature. In addition to Charles M. Child (1869-1954), Charles J. Herrick (1869-1960), and Karl S. Lashley (1890-1958), of whom he already knew some works, John Dewey (1859-1952) and Frank R. Lillie (1870-1947) were illustrious representatives of this tradition, sustained by the politics of its president (Robert M. Hutchins). It manifested itself in a strong interest in a holistic approach to the living organism, to the mind and to society, which created a very favourable climate for interdisciplinary work and supported the effort to develop an approach to social phenomena rooted in biological sciences.

Especially from late 1937 onwards, Bertalanffy met the neuro-physiologist Ralph W. Gerard (1900-1974), whose work, like that of Lotka by the way, was inspired by Herbert Spencer (1820-1903) and was then oriented towards an organicist approach to the general problem of the transmission of messages on all levels of organization, among them the social level.

5.1.3 The First Presentation of a General Systemology

Even in the United States Bertalanffy remained in contact with Austrian and German neo-positivism, of which a small group began to form precisely at this time in Chicago under the auspices of the pragmatist philosopher Charles Morris (1901-1979). In particular he met Carnap and Carl G. Hempel (1905-1997)[1] there again, who had just emigrated.

In a seminar organized by Morris in late 1937, where Bertalanffy was invited to give a lecture, he presented for the first time an outline of his project of a general systemology. In all sciences Bertalanffy noted the same promotion of a dynamist conception, of concepts of "wholeness" and "organization", as well as the recurrence of a formal identity of principles and laws in domains that are *a priori* completely different ("isomorphisms"). In order to explain this development, he postulated the existence of principles, of models and laws that

---

[1] Letter from L. von Bertalanffy to C.G. Hempel (06/11/1950), BCSSS archive.

apply to any type of system, regardless of the nature of its elements and its particular characteristics. Hence his project of a general systemology, whose goal was their formulation. It ideally had to reach the state of a logico-mathematical theory, capable of deducing *a priori*, from the abstract definition of a system and from the introduction of specific conditions, certain general laws about the systems with which not only the natural sciences, but also the humanities and social sciences are confronted. For Bertalanffy this "systemology" simultaneously had to be a logic and a methodology of systemic modelling, that had to facilitate everything by codifying the transfers of models between disciplines, avoiding superficial analogies by highlighting profound "homologies" and thus permitting the non physico-chemical sciences to progress towards a course of "accuracy". Its major purpose was in his eyes to generate a radically new type of unity of science, which stood in sharp contrast not only with an ideal of the nineteenth century that had still largely operated in sciences such as biology and psychology, but also with the physicalism advocated by one section of the Vienna Circle. It was not a unity based on the reduction of concepts, methods or even laws of all sciences to those of a single science considered more essential, such as physics, but a formal unity founded on the generality and ubiquity of the concept of the system and on the isomorphisms that it induced between the sciences, whose logical and methodological autonomy was guaranteed.

Unfortunately for Bertalanffy, his project was met with hostility. The reasons for this were certainly manifold, but they are unknown and can thus only be the subject of conjectures. It seems at least possible to suppose that Morris, who saw in semiotics, where he then laid the foundations for a meta-science that encompassed all individual sciences, had expressed a particularly vehement criticism concerning this project; all the more so as a conflict of interests matched him against Carnap within the University of Chicago: Bertalanffy appeared as the second thinker from Vienna to come and defend ideas that rival his own (Brauckmann, 1997: 6). Anyway, the criticism was such that Bertalanffy feared for his reputation as a scientist and decided to abandon at least provisionally the idea of defending his project publicly. This setback indicated also the beginning of an open hostility on his part towards positivism as a whole, whose intensity was only to increase thereafter.

5.2 Bertalanffy's Reaction to the Anschluss and the Rest of His Stay in the United States

In the course of his stay in Chicago, Bertalanffy was invited by several universities throughout the country, in particular in order to present his theory of organic growth, which was still in the progress of being elaborated. But he was limited in his travels by the fact that the Rockefeller foundation refused to subsidize them: the research in molecular genetics, which already absorbed 30% of his research budget, had general priority over the problem of organic growth (Brauckmann, 1997: 6).

Meanwhile events of a very different nature, which had serious consequences, occurred. Hitler annexed Austria on March 13, 1938, and Bertalanffy and his wife soon heard the news. Although worried about their son, who had stayed in Austria, they decided to continue their stay until its end. But Bertalanffy soon learned also of what was happening at the University of Vienna: several members of the *Vivarium* were dismissed from their posts on April 22 because they were Jews, in particular Przibram—who died in 1944 in the Theresienstadt concentration camp.

Bertalanffy left Chicago for good on May 4 towards Colorado and California, with a view to reconciling a leisure trip with lectures. He then had to join the laboratory of marine biology in Woods Hole (Massachusetts)

The Dialectical Tragedy of the Concept of Wholeness

in order to work there during the summer and until the end of the time allocated to him. Before his departure he wrote to Franck B. Hanson, director of the office responsible for natural sciences at the Rockefeller Foundation, so as to request an extension of one year of his grant—hoping to find a way so that his son could join them in America. He affirmed to Hanson that, although his family was not Jewish, he risked being deported to a concentration camp upon his return because he had made offensive remarks against the Nazis in public in the past. In fact he openly sought to obtain a post in an American university. The difficulty for him was that, despite the interest his work aroused, the Foundation was rather inclined to support researchers where it had been proved that they had been persecuted or were in the process of being persecuted, which was not the case with Bertalanffy, despite his statements. In early June he interrupted his journey in order to discuss the problem with Hanson personally. The latter informed him that the Foundation refused to extend his grant. Bertalanffy insisted that his case be reconsidered after consultations of "bio-mathematicians" such as Raymond Pearl (1879-1940) and Rashevsky. But Rashevsky, who met Weaver in this connection, did not support Bertalanffy and even assessed him negatively: their relative enmity obviously had a decisive influence at least as much on his judgement as on the purely scientific considerations. As a result, the renewal of the grant was finally refused (M. von Bertalanffy, 1973: 43-44; Brauckmann, 1997: 6-7).

When invited by the physiologist Selig Hecht (1892-1947), whom he already knew well for his work on the physiology of the senses that were founded on an approach which was ultimately based on the model of the open system, Bertalanffy left to give a lecture at the University of Columbia. After having been invited by Lashley (Brauckmann, 1997: 6-7), he gave another talk at Harvard, i.e., at the main seat of American "holistic" thought next to Chicago—besides Lashley since 1935,

Chapter 5: A First Trip to the US

Whitehead, Lawrence J. Henderson (1878-1942), and Walter B. Cannon (1871-1945) worked there.

Bertalanffy arrived finally at the laboratory of Woods Hole in late June and worked there until the beginning of October. He began a series of experiments on the growth of planarians, whose importance was "strategic" in his theory of organic growth. He also met Weiss there (Brauckmann, 1997: 6-7); but this reunion was probably not very friendly, as their relationship had deteriorated since the latter had claimed that Bertalanffy's "organismic" philosophy and concepts were an outright reiteration of his own (Hofer, 2000: 152)—something which, although caricatural, was not completely illegitimate as regards Bertalanffy's works before 1930. He also got to know the great geneticist Thomas H. Morgan (1866-1945) at Woods Hole, at a time when he just tried hard to integrate the advances in genetics and the general theory of evolution into his "organismic" philosophy. He also had the opportunity to talk to a famous critic of mathematical biology and, more specifically, of mathematical theories of growth (organic growth or growth of populations): Edmund B. Wilson (1856-1939) (M. von Bertalanffy, 1973: 44). But his most fruitful encounter at Woods Hole for his later work was without doubt that with the Canadian biologist A.C. Burton, who at the time worked like him on the problem of the dynamical equilibrium of open systems and contributed as much to his reflections in this regard as Reiner and Householder.

At the end of his stay at Woods Hole, Bertalanffy and his wife departed on a journey back to Austria. He embarked on October 8, 1938, at New York to return to Vienna, after Ludwig had declined the offer of a colleague from Chicago, who had proposed to accommodate him and to arrange the continuation of his career in the United States (Davidson, 1983: 55).

CHAPTER 6
Bertalanffy as a Biologist of the Third Reich (1938-1945)

6.1 Bertalanffy's Membership of the NSDAP (1938): His Reasons and His Benefits
6.1.1 A Return to Vienna in a Favorable Conjunction of Circumstances

Bertalanffy had hardly returned to Vienna when Versluys, still head of the Department of Zoology, proposed in late October that he be given a course of three hours per week in order to compensate for the absences of Przibram and Andreas Penners, who had been dismissed from their posts. Versluys also wanted Bertalanffy to set up "exercises in experimental zoology" as a replacement of those previously conducted by Przibram and to be paid accordingly. He in fact had the intention that Bertalanffy should sooner or later take over Przibram's chair. The initiative of these requests can probably be attributed to Max H. Knoll (1897-1969), then president of the committee of the Academy of Sciences, who knew Bertalanffy's work precisely and particularly appreciated its development into the direction of biophysics. Versluys's request was acceded to by the university administration. On November 18, Versluys also submitted a second request, demanding that Bertalanffy be appointed from December 1 onwards as the temporary replacement of an assistant professor of the Department of Zoology, called Schremmer, who had been called to arms (Hofer, 1996: 18-19; Brauckmann, 1997: 8). He was to have this status from December 1, 1938, to September 30, 1939.

Bertalanffy had even less reason to complain about his fate, as he, equally in this period, received (and accepted) the prestigious offer from a German publisher to organize the publication of a collective biological work of encyclopaedic character, the *Handbuch der Biologie*: for that purpose he was supposed to coordinate the work of about fifty recognized German and Austrian biologists.

6.1.2 The Membership in the NSDAP

Bertalanffy's return to Vienna occurred thus in a favorable conjunction of circumstances. Nevertheless, in his view this was not synonymous with a rapid and secured promotion to an academic position worthy of the name. Versluys's requests, even though they went into this direction, did by the way not yet ask for a fixed post and were not in accordance with the idea Bertalanffy, then 37 years old, himself had of the position, which should finally go to him because of his qualities.

He decided on November 20, 1938, to submit an application to join the National Socialist Party (NSDAP). As a reason he gave his contribution of his biological writings to the scientific development of the National-Socialist vision of the world. And he presented himself as a victim of the regime before the *Anschluss*, claiming that his sympathies for the National-Socialist movement (then conflicting with the politics of the government, which first suppressed this movement before it could merely try to contain it) had been the cause of his professional stagnation and his financial difficulties.

The official responsible for endorsing the applications did not only respond positively to the request: he noted that he had known Bertalanffy's sympathies for National Socialism even before the annexation of Austria, and congratulated him on having made donations to the movement despite his difficult financial situation (Hofer, 1996: 20-21). Such allegations remain doubtful, considering a context where the organization of a lie is cultivated like an art. Bertalanffy's entry into the NSDAP poses nonetheless the problem of his ulterior motive and, more generally, that of his relation to National Socialism: is there more behind this membership than opportunism?

6.1.3 The Complexity of Bertalanffy's Relation to National Socialism

The first fact that needs to be considered in order to respond to this question is that Bertalanffy, before Hitler came to power in January 1933, had repeatedly defended positions that were clearly opposed to a certain number of National-Socialist theses and, to a larger extent, to some general aspects of the ideological and cultural context favourable for its advent. We have thus seen that he vigorously criticised the ideological instrumentalization of biology in 1930, in particular of Darwinism, which was the case until 1932. His demonstration of the insufficiency of the scientific foundations of "racial hygiene" and eugenics passed through a critique of some of the major ideologists of National Socialism such as Eugen Fischer (1874-1967), Fritz Lenz (1887-1976), and Hans F.K. Günther (1891-1968). Until 1932 Bertalanffy also attacked the irrationalist drifts of "a 'philosophy of life' that is not always delightful" (L. von Bertalanffy, 1932: 4), as well as the "ghosts and grimaces of a mysticism" whose aim is to bring modern times to the "barbaric forms of medieval Christianity", to "the night of ignorance and faith"[1]—important elements in Hitler's rise to power.

It is actually quite clear that Bertalanffy did not embrace certain basic patterns of National-Socialist ideology and scorned a great deal of the ideas, slogans, and actions that made it successful. But this does not make

---

1. L. von Bertalanffy, 1927: 264. Translated from the German by ES. See also L. von Bertalanffy, 1930: 46-50.

him a thinker who was principally opposed to all aspects of this ideology, even before 1933. Bertalanffy was a professed conservative, who was convinced of living in a decadent period and who already at the age of 23 believed that "our task, the task of a diminishing sort, is not to invent new values, but to faithfully preserve the old ones".[2] While affirming the need to resist to the nihilist desire to "throw over-board" our Western civilization and its intellectual creations even if it is "old and tired", he clearly fit into, with his recurrent attacks against a "mechanized", "materialist" and "soulless" world in which man would be enslaved by physical technology,[3] an anti-modernist movement that had been very important in the German-speaking academic world since the end of the nineteenth century. No less important and generally linked to this anti-modernism were furthermore the anti-democratic, anti-liberal and anti-socialist convictions that Bertalanffy shared. These beliefs, particularly evident in his post-war writings and correspondence, were already apparent before 1933: his critique of Lamarckism and environmentalism was simultaneously that of socialism; his critique of Darwinism and its reincarnations was also that of liberalism and bourgeois values, even more than that of racist and eugenic doctrines. As for his hostility towards the democratic system, it was an expression of an aristocratism that was never to abate, and was underlying his critique of an egalitarian educational policy, his adherence to theses that reduced democracy to a "slogan that serves an underground, absolutist financial power", or even his regular critique of an "atomistic" conception of society conceived as an amorphous sum of interchangeable individuals.[4] Bertalanffy thus shared political and ideologi-

---

2. L. von Bertalanffy, 1924b: 343. Translated from the German by ES.

3. In particular L. von Bertalanffy, 1926a and L. von Bertalanffy, 1928: 229-230. Translated from the German by ES.

4. L. von Bertalanffy, 1924a, III and IV; L. von Bertalanffy,

Chapter 6: Bertalanffy as a Biologist of the Third Reich

cal positions that, without being National-Socialist in themselves, constituted a common fund which Hitler's movement itself fed on and thanks to which it prospered; a common fund which explains to a large extent the attitude towards this predominantly passive and often uncritical movement of an academic world of which relatively few members actually subscribed to the characteristic theses of National Socialism.

It is therefore understandable that Bertalanffy, without sharing these, cultivated even before 1938 a lasting friendship with Alfred von Auersperg (1899-1968), who became an S.S. physician after the *Anschluss* and about whose National-Socialist beliefs there is no doubt. His relationship with this neurologist of the University of Vienna, appointed "associate professor" in 1943, continued even after the war, when Auersperg took refuge in Chile.[5] Several of Bertalanffy's friends, whether out of opportunism, out of conviction, or both, collaborated in fact with the Nazis. Such was even the case with the best among them, Gessner,[6] who also maintained a lasting friendship with Auersperg and planned in 1950, just like Bertalanffy himself a little later, to join him in Chile.[7]

A particularly important moment as regards Bertalanffy's relation to National Socialism, because it is perhaps the most revealing, occurred long before 1938. It goes back to an article published in 1934. Before that, his critical reflection on the category of "wholeness" and the concept of an organization of "superior order" were

---

1930: 35-43; L. von Bertalanffy, 1932: 4-5. Translated from the German by ES.
5. Correspondence between L. von Bertalanffy and A. von Auersperg (15/05/1950; 01/09/1950; 10/05/1952; 15/05/1952), BCSSS archive.
6. Letter from F. Gessner to L. von Bertalanffy (06/12/1946), BCSSS archive.
7. Letter from F. Gessner to L. von Bertalanffy (23/04/1950). Letter from A. von Auersperg to L. von Bertalanffy (15/05/1950), BCSSS archive.

essential. It led him on the one hand to being in various respects on the same wavelength, since he applied his critique of "atomistic" conceptions to sociology, with the dichotomy between "community" [*Gemeinschaft*] and "society" [*Gesellschaft*] that was then fashionable and came from Ferdinand Tönnies (1855-1936); the first, seen in a positive light, was assumed to be an "organic whole", while the second, perceived negatively, was considered equivalent to a mechanical entity. One finds this dichotomy also in the older one between "culture" and "civilization", equally made fashionable again by Spengler. Now these dichotomies also contributed to the success of National Socialism, since they were used for legitimizing a mystique of the people that, precisely, was embodied among other things in the concept and slogan of the *Volksgemeinschaft* ("community of the people"). Bertalanffy could have contributed directly to this legitimization from 1926 onwards. But instead he repeatedly attacked all forms of a mystique of "wholeness" by stressing its dangers, given that his "organismic" philosophy specifically aimed at paving the way for a scientification and a demystification of the categories of "wholeness" and organization. And his fascination for Spengler did not prevent him from criticizing, from 1924 onwards, the reification of his organicist model of cultures by denouncing his incomprehension of Goethe (whom Spengler claimed to follow) and of the morphological concept of the type, and to renew this critique in his thesis by denouncing the blurred boundaries between "realities" and "mere conceptual models".[8] As to biologism, Bertalanffy rejected it explicitly in the last part of his doctoral thesis. There is thus an inherent ambiguity in Bertalanffy's "organismic" thought. It is very revealing that in the legitimization of National Socialism it only changes after

---

8. L. von Bertalanffy, 1930: 47-50; L. von Bertalanffy, 1924a, III; L. von Bertalanffy, 1926b: 87. Translated from the German by ES.

Chapter 6: Bertalanffy as a Biologist of the Third Reich

1933, in a German magazine—which had then been condemned in Austria, and it is furthermore likely that this change was linked to the contemporary efforts of Jordan to enable him to enjoy once again the (German) research scholarship, which he had received until Hitler's accession to power:

The organism appears no longer, as earlier in the theory of the "cell state", as a republic of autonomous parts with the same rights, but rather like a hierarchical structure, dominated on each level by the Führer *principle.*[9]

This ambiguity can be found in an almost caricatural manner after the war, when Bertalanffy, while also strongly criticizing the organic analogy in sociology, corresponds with the anthropologist and philosopher of history Friedrich Keiter (1905-1967), a former Austrian member of the NSDAP and great theorist of a "biology of culture" [Biologie der Kultur] and racial hygiene.[10] It should also be noted that many representatives of the philosophies of "wholeness" that were linked to Bertalanffy by recurrent mutual references, such as Alverdes, Jakob von Uexküll (1864-1944), Bavink, and Adolf Meyer-Abich (1893-1971), did not hesitate to actively invest in the ideological support of the National-Socialist regime from 1933 onwards, or even earlier.

Later developments showed, however, that Bertalanffy's own contributions to the legitimization of the latter through his "organismic" philosophy were in fact very limited and were systematically put forward in a

---

9. L. von Bertalanffy, 1934: 352: "So erscheint der Organismus nicht mehr wie früher in der Lehre vom 'Zellenstaat' als eine Republik von gleichberechtigten und gegeneinander selbstständigen Teilen, sondern weit eher als sein hierarchischer, in jeder Stufe vom Führerprinzip beherrschter Aufbau". Translated from the German by ES.
10. Correspondence between L. von Bertalanffy and F. Keiter (19/01/1949 – 04/04/1950), BCSSS archive.

manner so as to advance his own work; and that beyond these few assertions there was no contribution on his part to the "*oeuvre*" of the Hitler regime.

All these considerations seem to lead us to the conclusion that Bertalanffy, who entered the NSDAP of his own free will, did it above all out of opportunism with the hope of giving a powerful impetus to his career and of finally obtaining a position at the university which he had long coveted in vain. It is likely that if he had already acquired a satisfactory position before 1938, he would have adopted the majority position of, possibly uncritical, passivity mentioned above, without feeling the need to join the NSDAP: the decisive factors are evidently the resentment connected to the marginality of his position and the use of a favorable conjunction of circumstances which he deemed opportune to exploit as fast as possible. Something which, let us note well, is not really justified, as the joining of the party was not at all a guarantee—a fact that is known by historians today, but was probably not known by Bertalanffy.

6.2 The Promotion to the Title of "Associate Professor" (1938-1940)

A new element in view of a genuine promotion of Bertalanffy occurred on 24 December, 1938: the University of Vienna sent a request to the Ministry of Science and Education in Berlin in order that a post of a [*außerordentlicher*] professor of theoretical biology, next to the already existing posts in the department of zoology, would be created. The subsequent correspondence shows that the position in question was explicitly intended for Bertalanffy, and it is also likely that Knoll exerted his influence on the initiative of this request. However, the request was rejected.

6.2.1 A First Step: The Promotion to the Position of a Lecturer with Civil-Servant Status (1939)

Bertalanffy thus had to content himself with a teaching post as an assistant next to his research until September 1939. He taught an integrated course of zoology and botany for medical students—and was the first biologist to offer such courses. But his salary was small and his situation precarious, so that he tried hard to improve his situation. On 21 April 1939 he sent a request to the dean of the Faculty of Philosophy for the promotion to the rank of "associate professor". As reasons he gave the instability of his status, his low income with which he had to support wife and child, and the fact that his situation would deteriorate once Schremmer would be reinstated in his post. In addition he claimed that he had gone to America in 1937 because of the hostility towards his National-Socialist orientation and the professional dead end that it involved, and that he had come back voluntarily in order to serve the *Reich* once the annexation of Austria had been completed. But three months later, the dean had still not responded to the request.

So Knoll intervened. On 21 July he sent a request to the Ministry from the Faculty of Philosophy in order that Bertalanffy should get the status of a lecturer together with that of a civil servant as a temporary solution, while waiting for the post of a full professor. The request succeeded on 27 October, 1939, when Bertalanffy was charged to offer courses in theoretical biology. And he pledged allegiance to the *Führer* as a civil servant of the *Reich* (Hofer, 1996: 21-23; Brauckmann, 1997: 7-8) on 8 December 1939.

6.2.2 Bertalanffy as an Associate Professor: Some First Moves

Bertalanffy had hardly been appointed, when he reiterated his application for a post as an associate professor. On 15 February 1940 he filed a request to that effect with strong support from the director of the Institute of Zoology (H. Weber). In his letter of recommendation Weber praised Bertalanffy's scientific qualities and his recognition on the international level; but he also insisted on the professional obstacle constituted by the so-called political beliefs before the *Anschluss* and pointed out that his membership in the NSDAP qualified him for the post. The dean's report to the president of the board of professors at the University of Vienna, dated February 20, emphasized that Bertalanffy was scientifically and ideologically able to take the post and "strongly" recommended his nomination. The request was accepted on 20 September 1940: Bertalanffy was appointed associate professor of general biology and experimental zoology (Hofer, 1996: 23-25; Brauckmann, 1997: 8).

He also managed to get the necessary support in order to avoid being called to arms, even as several of his colleagues did not enjoy the same favour.

During this period Bertalanffy initiated also an administrative procedure that aimed at having recognized his seniority as a lecturer as well as the "fact" that his status since 1930 had suffered because of his political views: he demanded that his years of service should be retroactively revaluated equivalent to his years as a lecturer with the status as a civil servant and that the corresponding (substantial) financial compensation should be paid to him. The request received again Knoll's support and was partially accepted (a period of four years and three months was withheld; Hofer, 1996: 24-26).

Chapter 6: Bertalanffy as a Biologist of the Third Reich

6.2.3 Bertalanffy's Work between 1938 and 1940: An Outline of a "Theory of Open Systems" and First Impacts of the Political Context

Bertalanffy produced only five publications between his return to Vienna in late 1938 and late 1940. The context, the teaching load, and the concerns raised above were not conducive to serene and prolific research. Three articles concerned his theory of organic growth. One, published in late 1938 in the United States, was essentially only a translation of his fundamental article of 1934. As to the others, they concerned his recent experiments on the growth of planarians, organisms that allowed him to show at the same time the relevance of the hypotheses of his theory and the correspondence of his mathematical predictions with the experiments.

One of Bertalanffy's publications during this period, in August 1940, constituted one of the most important of his career. Entitled "Der Organismus als physikalisches System betrachtet" ("The Organism Considered as a Physical System"), it appeared largely as the result of his exchanges with his American colleagues from 1937 to 1938. In this article he outlines a general theory of open systems, purely formal and thus *a priori* applicable not only to biology, but also to chemistry, as well as to demographical and sociological problems. Bertalanffy shows in particular the insufficiency of classical thermodynamics for the theory of biology and the fact that equifinality, impossible in systems that are closed to exchanges of components with their environment, is a necessary and general characteristic of open systems insofar as they attain a dynamic "pseudo-equilibrium". But, limited in the elaboration of this theory at the same time by the absence of an adequate thermodynamics and by the formalism applied, he needs to resort to the construction of a model, in the form of a hypothetical (chemical) open system, in order to illustrate the close link that he also tries to establish between the openness of systems and their capacity for self-regulation and adaptation. This article plays a central role in his work. On the one hand, the development of such a theory of open system fits perfectly with the continuity of his "organismic" programme in biology, which he had called for since 1932, and which had become all the more established ever since Bertalanffy had asserted that the openness of organized systems is their primary characteristic. But on the other hand, it also responds to the need for elaborating his project of a general systemology, presented without success in 1937, more precisely to provide it with a "concrete" basis of legitimization that it still lacked. In fact, this theory of open systems amounts largely to an exposition of a general systemology in a specific form that allows Bertalanffy to demonstrate its feasibility and potential productivity.

Still in 1940, Bertalanffy published his sixth book, *Vom Molekül zur Organismenwelt* ("From the Molecule to the World of Organisms"). This short work, which consists essentially of a non-technical exposition of the theses proposed in *Das Gefüge des Lebens*, is notably characterized by the introduction of the term *Fliessgleichgewicht* (steady state) for referring to the specific type of the "dynamic (pseudo)equilibrium" that occurs in open systems. But it is also marked in various ways by the political context. Bertalanffy brings about a softening (necessarily for the occasion) of his critique regarding evolutionary theories of Darwinian inspiration. Genetics plays an even more important role than in *Das Gefüge des Lebens*: it becomes a major topic of discussion—in the "organismic" perspective. What is more, by applying it more particularly to humans, Bertalanffy strives to link his theory of organic growth with diverse anthropological theories that were in fashion and were largely used in an anti-Semite perspective, also known as the "types of human constitution" [menschliche Konstitutionstypen]. He links it in particular to that of Ernst Kretschmer (1888-1964), who distinguishes such biological "types" by correlating them to psychological characteristics and psychiatric determinants. Another aspect of the political context in this book is his clear reflection on a fundamental moment in the development of the Third Reich, which occurred in 1938 and 1939. This moment consisted in a reorientation of the conception of the role of science: while it was first essentially a tool for ideologically legitimizing the regime, it became more and more subject to utilitarian imperatives—imperatives that were diametrically opposed to the idea that Bertalanffy made of it. In this work one can therefore also see him on the one hand insist on the practical usefulness of the "organismic" conception, in particular as regards ecological problems and demographical policies, and on the other hand, pointing out that it would be a serious mistake to orient research only towards what can have immediate practical applications.

6.3 Bertalanffy as an Academic of the Third Reich (1940-1944)
6.3.1 The General Theory of Organic Growth, Beginnings of a "Dynamic Morphology"

Having obtained his post as "associate professor" in September 1940, Bertalanffy found himself finally in conditions that were favorable for advancing his research on the connections between metabolism and organic growth, in particular on the relative growth of the organs. He published four articles on the topic in 1941, the last of which was his mathematical theory of global organic growth. It provided a general form of equations for growth that applied to the whole kingdom of animals and established a connection between "types of growth" and "types of metabolism".

During this period Bertalanffy also completed the second volume of his „theoretical biology", exclusively devoted to problems of the metabolism and of growth.

This volume was published in early 1942. He justifies there the systematic application of the mathematical analysis to his problems and implements it. In particular he formulates a synthesis between his theory of global organic growth and the law of "allometric" growth that governs numerous problems of relative growth, outlining thus a general mathematical theory of organic growth. This theory constituted for Bertalanffy the draft of a „dynamic morphology", based on a conception of the organism as an „expression of an ordered flow of events", where what persists is not a fixed structure, but the systemic order of the underlying processes, the legality that expresses them and that needs to be determined. The ultimate aim of this "dynamic morphology" was to engender a unification of seemingly diverse domains such as morphology, physiology, genetics and the theory of evolution.

In this work Bertalanffy, who did not only draw on Lotka and Volterra, but also on Pearl and Vladimir A. Kostitzin (1883-1963), from then on also for the first time took into consideration in detail the problem of growth considered from the point of view of population dynamics. He appropriated in particular the project of general kinetics of Lotka's dynamic systems and developed, in an extension of his article on open systems of 1940, the basics of the material of his subsequent articles on the general systemology. He thus also took up Lotka's work again, which consisted in discussing *a priori* the properties of steady states of dynamic systems and in formally deriving "laws" of growth that can be applied to the most diverse scientific domains: exponential and "logistic" "laws".

6.3.2 Bertalanffy's Explicit National-Socialist Commitment in his Writings in 1941

But Bertalanffy's concerns were far from being limited to a "purely" scientific framework. And the following year,

Chapter 6: Bertalanffy as a Biologist of the Third Reich

1941, marked the culmination of Bertalanffy's National-Socialist commitment in the form of an article published in the journal *Der Biologe* ("The Biologist"). This journal was the official organ of National-Socialist biology and was placed directly under the supervision of Heinrich Himmler (1900-1945). In this article Bertalanffy tries to present "organismic" biology both as an expression and as a scientific justification of the National-Socialist vision of the world. He explicitly links his concept of hierarchical order to fascism, welcoming a contemporary rupture with "atomistic conceptions of the state and of society" in favour of a "biological" conception that "recognizes the wholeness of life and the people" and "satisfies the hope" of such a rupture, which he advocated from 1928 onwards in his *Kritische Theorie der Formbildung*.[11]

Some racist reflections appear equally in this text. Bertalanffy, under the influence of Spengler, of Uexküll, and of several neo-Kantian traditions, had defended a "perspectivist" philosophy of knowledge based on the idea of a "biological and cultural relativity of categories", as he would call it in 1955, since the 1920s (more explicitly since *Das Gefüge des Lebens*). But one can see him write here that the development of science is conditioned by the psychophysical organization of man, and in particular by "racial dispositions", and that "the scientific vision of the world represents a specific product of the Nordic spirit". The same kind of reflection can be found in the same year (1941) in another article: Bertalanffy writes that "the primitive races have the same cephalization as the man of culture", but that "they do not use it fully".[12] One needs to note, however, that these are the only racist statements made by Bertalanffy in the course of his whole career.

---

11. L. von Bertalanffy, 1941b: 341, 343. Translated from the German by ES
12. L. von Bertalanffy, 1941b: 337 and L. von Bertalanffy, 1941a: 16. Translated from the German by ES

Whether speaking of these statements or of his more general attempt at unifying his "organismic" philosophy with National Socialism in a process of mutual justification, it is, moreover, quite clear that one needs to understand Bertalanffy's involvement not as an expression of profound convictions, but as an effort aimed at making his work known in the context of an increasingly unfavourable relation of the National-Socialist regime with science in this regard. It is indeed symptomatic that in *Der Biologe* Bertalanffy redoubles his efforts, more than in 1940, for arguing the essential importance of theoretical work and of the determination of "exact" biological laws in view of practical ends—taking the theory of heredity as a model. The principal objective of his article was to demonstrate the productivity of "organismic" biology in this perspective.

6.3.3 The Art of "Manipulation" in order to Achieve One's Ends

In order to demonstrate the relevance of his "organismic" philosophy in the context of the Third Reich, Bertalanffy used a certain number of "manipulations" from late 1941 onwards, in which important people supported him, that were meant to prepare his promotion to the rank of professorship, to maximize his salary, as well as to protect him from the inconveniences related to the state of war.

In autumn 1941 Weber was called to take a post in Strasbourg. As he had offered some of his courses at the Faculty of Medicine, the Faculty of Philosophy was asked whether Bertalanffy, who had already taught such courses in 1939, would fill in for Weber while being assured of keeping this teaching on the return of the latter. The request was only partially satisfied, after a conflict between the two faculties. The course "Zoology for Physicians" was only allocated to Bertalanffy for the first semester of 1941/1942.

Chapter 6: Bertalanffy as a Biologist of the Third Reich

Bertalanffy had not given up at the same time to obtain a more substantial re-evaluation of his years of service as a civil servant at the University of Vienna. He adopted a new approach in March 1942. He demanded even then that all of his past years since obtaining his doctorate in 1926 should be taken into consideration. He justified himself by claiming that his work between 1926 and 1934 was aimed at integrating the National-Socialist ideology in an important scientific domain, which had become essential in modern biology and of which one had to consider him as a leader. His allegations were supported without reservation by Knoll, who considered that he deserved this retroactive recognition scientifically as well as ideologically. The request was again largely acceded to in June 1942 and covered the entire period from 1 May, 1931 to 1 April, 1940 (Hofer, 1996: 27-28).

In April 1942 Bertalanffy received a call to military service. He then appealed to Knoll, who did his best with the competent authorities, claiming that Bertalanffy was indispensable because he was the only zoologist that had not yet been summoned to military service and that his teaching at the Faculty of Medicine performed a useful mission in times of war. He was even supported by the editor of the *Handbuch der Biologie*, who put forward that Bertalanffy, irreplaceable in organizing the encyclopaedic work, performed the mission of defending the scientific pre-eminence of Germany against the rising power of America. Finally, he won the decisive support of *Reichsleiter* Baldur von Schirach (1907-1974)—who was to be judged and sentenced to twenty years of prison in the Nuremberg trials. But the exemption was short-lived: it was only valuable until 30 June, 1942. The argument that Bertalanffy was irreplaceable was weakened by the nomination of Wolfgang von Buddenbrook (1884-1964) as head of the Department of Zoology in June. Steps in the same direction were thus undertaken, with a new argument, namely that Bertalanffy assumed the important task of teaching the theory of heredity, due to which he remained irreplaceable. And he got once again a provisional exemption (Hofer, 1996: 29-31).

Bertalanffy did not intend to content himself with the post as associate professor either. On 17 July 1942, a request of the Department of Zoology was sent to the Ministry in Berlin, without being passed on to the dean of the Faculty of Philosophy, demanding that a chair (*planmäßiger Professor*) in "theoretical and quantitative biology" should be created for Bertalanffy. In the *curriculum vitae* that he attached to this demand, the latter emphasized not only the recognized quality of his works in the mentioned fields, but also the "fact" that they "correspond fully to the aims pursued by the Third Reich".[13] No action followed this request, however, and the dossier remained unanswered at the Ministry until 1948 (Brauckmann, 1997: 8).

6.3.4 Bertalanffy's Scientific Activities between 1942 and 1944

From the scientific point of view, the years 1942 and 1943 were productive for Bertalanffy. His research was almost entirely focussed on the problem of organic growth, on which he published six articles. But his writings were mostly limited to putting his theory to the test of experiments and basically did not develop its foundations at all.

In one of them, published in 1943 in the *Zeitschrift für Rassenkunde* ("Journal for the Knowledge of Races"), Bertalanffy insisted once again on the possibility of linking his theory of organic growth to the question of human "types of constitution". He planned in fact in this period a research programme, founded on this theory and dealing with the relation between these "types" and metabolism. He had several of his students work on the application of D'Arcy W. Thompson's (1860-1948) "method of

---

13. Hofer, 1996: 28-29. Translated from the German by ES.

transformation" on human phylogeny. But he ultimately abandoned this programme due to a "lack of experimental material". This programme, whose ideological motivation removed nothing of its interest from an anthropological perspective, was to be taken up by others after the war on the basis of his work.[14]

In 1943 Bertalanffy was equally interested in "basic biological units", genes and viruses, which were then subject of active research for not only ideological, but above all economic and sanitary, reasons. In an article published in January 1944, he expounded a model of such basic units as "one-dimensional metabolic crystals". A common quality of these units, their capacity of "covariant reduplication", appeared as a consequence of the "fact" that they constituted open systems. Although this model was soon revealed as inadequate, it constituted a remarkable attempt at extending the "organismic" point of view in a field that was soon to triumph and to constitute a most troublesome "competitor": molecular biology.[15]

In late 1943, Bertalanffy furthermore started a correspondence with the German philosopher Erich Rothacker (1888-1965), who planned to write a philosophical history of biology similar to that outlined by Ungerer in 1942 in the first volume of the *Handbuch der Biologie*. And he wanted Bertalanffy to write the chapter on general biology. But the correspondence did not succeed and this contribution, which would again be mentioned after the war, was not written (Brauckmann, 1997: 8-9).

The year 1944 was only seemingly little productive: there was no article published after January. But on the one hand, Bertalanffy put a lot of energy into his course for the medical students, where he advocated

---

14. Letter from L. von Bertalanffy to W. Selberg (05/10/1951); letter from L. von Bertalanffy to D. Stark (26/06/1952), BCSSS archive.
15. L. von Bertalanffy, 1944. Translated from the German by ES.

and outlined an "organismic" medicine in which the concepts of the primary activity of the organism and of psychophysical unity played a central role. This course supplied him with the material for two short books that he wrote in 1945: *Biologie für Mediziner* ("Biology for Physicians") and *Biologie und Medizin* ("Biology and Medicine"). On the other hand, Bertalanffy worked on the third volume of his *Theoretische Biologie*, whose contents remain unknown. Finally, he resumed his project of a general systemology and wrote a fundamental article for one of the leading German journals of philosophy (*Deutsche Zeitschrift für Philosophie*). The article was sent to the journal in late 1944 and the proofs were returned to him for correction on 6 February 1945. But it never appeared, as the publication of the journal in question became impossible in the torments of the end of the war.

6.3.5 The Impact of Bertalanffy's Work during the War

Bertalanffy's writings, although essentially those that were written before that period, were commented on and used several times in the German-speaking world between 1940 and 1944. His concept of the organism as an open system in "dynamical pseudo-equilibrium" inspired, for example, largely and favourably the reflections of the biologist Heinz Dotterweich (1904-1949), who published a general study on the concept of the equilibrium in biology in 1940. The philosopher Theodor Ballauf (1911-1995) demonstrated, for his part, in two articles in 1940 and 1943 the remarkable parallel between his "organismic" philosophy and the "theory of categories" [Kategorienlehre] of N. Hartmann. As regards the Viennese psychologist Norbert Thumb, a student of Karl Bühler, he published an article aimed at showing the relevance of Bertalanffy's biological conceptions for his discipline in 1944.

On the other hand, there was obviously little response to Bertalanffy's writings during the war. An exception is his article from 1940 on open systems, which in particular in 1942 inspired the work of two German bio-physicists (U. Dehlinger and E. Wertz) oriented towards the development of a physical model of genes and viruses—a work that in turn influenced Bertalanffy's article on the subject in 1944. By contrast, his theory of organic growth was fairly ignored outside a very small circle of biologists, in a context where such a research field was, it is true, far from being privileged.

It appears in fact that the scientific, philosophical, ideological, and more widely, cultural context, which had been very favorable before 1933 so that his „organismic" conceptions could attract great interest, later gradually changed in a contrary direction, above all from 1939 onwards. Consequently one understands Bertalanffy's efforts, recurrent from 1940 onwards, to highlight the ideological and practical interest of his work and, simply, to exist on an academic level.

6.3.6 A Privileged Situation

Nonetheless, the fact remains that Bertalanffy benefited largely from the context of the Third Reich, whether in terms of his career or else concerning his financial and material situation.

An illustration of his privileged situation can be found in late September 1943. Bertalanffy received then another call to military service that he could not avoid, but whose duration was limited until December. From that moment on until the end of the war, besides the arguments already mentioned in this regard, he used that of an allegedly limited physical fitness for active military service. And indeed, Bertalanffy never again had to do military service: he was the *only* one to have such a privilege in the Department of Zoology in Vienna. Besides, his son Felix, first legally protected by his medical studies,

6.4 The Family Disaster at the End of the War

But the war ended dramatically for Bertalanffy and his family, brutally disrupting this situation. In March 1945, when his state of physical and above all moral health was very bad,[16] he left his residence before the siege of Vienna by the Soviets and took refuge with his family at his grandmother's home. The street battles, and above all the "scorched earth policy" practised by the SS soldiers, destroyed the district where their home was situated in April. When the siege was over, they found little more than ashes at their return. Nothing of their movable property or their clothing was left. But, much more seriously, Bertalanffy's library, which contained no fewer than six thousand books and eight thousand articles, among which many old scientific works, had entirely gone up in smoke. The same is true for three almost finished manuscripts (among them the third volume of his *Theoretische Biologie*), records of his experiments, his correspondence, his valuable stamp collection (collected since his youth with great erudition), his not less valuable collection of architectural drawings of the Renaissance and also other works of art.[17]

---

16. Letter from F. Gessner to L. von Bertalanffy (08/09/1949), BCSSS archive.
17. Letter from L. von Bertalanffy to J. Alexander (11/12/1950); letter from L. von Bertalanffy to S.T. Bornemisza (07/04/1951); letter from L. von Bertalanffy to V.A. Kostitzin (12/04/1949); letter from L. von Bertalanffy to W.W. Nowinski (30/01/1950), BCSSS archive. M. von Bertalanffy, 1973: 44-45; Brauckmann, 1997: 9.

CHAPTER 7
His Last Years in Vienna: The "Denazification" and Its Consequences (1945-1948)

7.1 The Period Immediately after the War: Hope of Continuity

Bertalanffy, the only one among the professors and lecturers of the Department of Zoology who was still in Vienna, was designated its provisional director by the dean of the university on 17 April, 1945. Charged with the examination of what could be saved of the scientific material at the university, he discovered a field of ruins. Almost all the material, documents, and equipment, were burned or unusable. Nothing was left of Bertalanffy's office or of his laboratory. More generally, two thirds of the department had been destroyed by bombing. Bertalanffy decided to work at its restoration so that the courses and the research could be taken up again as quickly as possible. Helped by three research assistants, he reached this goal at least partially within six weeks, so that the courses could resume in June—with almost eight hundred students enrolled in his classes. He decided furthermore to take the responsibility of reviving the Viennese journal *Biologia Generalis*, to which he had already contributed before the war (M. von Bertalanffy, 1973: 44-45; Brauckmann, 1997: 9).

He also took the opportunity of circumstances that were apparently favourable for reiterating a request for promotion, this time in order to obtain the highest position—namely, the chair of the Department of Zoology. A letter in this direction, which he had probably written himself, was sent to the dean of the Faculty of Philosophy on 9 July. This letter explicitly asks for the creation of a chair of general biology that should be awarded to Bertalanffy. The letter emphasizes the international recognition of his work and its transdisciplinary productivity; but also his "action for the new Austria", his resumption of *Biologia Generalis* and his partial restoration of the Department of Zoology are put forward in this sense. It also claims (which is particularly daring) that his career had been "strongly impeded" by the Nazi regime and that his status had consequently never corresponded to his scientific rank, since his demand for obtaining a professorship had been rejected in 1937 "for political reasons (anti-nationalist opinions, citation of Jewish authors, conflicts with the Ministry in Berlin because of his Austrian nationality)"[1].

Bertalanffy, who had obviously no doubts about his future, thought also in the course of this period about founding an *Institute for Advanced Studies in Biology* in Vienna with the physicists Arthur March (1891-1957) and Erwin Schrödinger (1887-1961). And he soon began negotiations for that purpose with the UNESCO But the circumstances were not favorable: he lacked support and the project was dropped—he later invoked "intrigues" to explain this failure.[2]

7.2 The Procedure of "Denazification"
7.2.1 A Failed Attempt at Escaping the Procedure

In fact, Bertalanffy had to pass, before anything else, through the commission of "denazification" in order to clarify his activities under the Third Reich. He

---

1. Hofer, 1996: 32-33. Translated from the German by ES.
2. Confidential letter from L. von Bertalanffy to the Ford Foundation (20/06/1958); letter from L. von Bertalanffy to A. Butenandt (19/01/1960), BCSSS archive.

Chapter 7: His Last Years in Vienna

knew it from the outset and took precautions even before sending his letter to the dean demanding his promotion: he left a request for exemption from registration before the chancellor and before the mayor of Vienna as a former member of the NSDAP (Hofer, 1996: 34). In order to justify his membership in the party, which he falsely dated at 1940, he claimed from May 1945 onwards in all his letters relating to it that he was of mixed race in the second degree because his maternal grandmother was Jewish; and that he could consequently only keep his place at the university by avoiding an examination that aimed at proving his Aryan lineage, which only his entry into the party could permit him. In reality, the available documents do not only show that Bertalanffy was by no means of mixed race in the second degree (what he himself wrote to Hanson in 1938), but that he was in fact also subjected to an examination that proved his Aryan lineage and whose result was positive in 1939 (Hofer, 1996: 20-21).

Without these elements being known, he did not escape a first suspension from his post in August 1945. But the mayor of Vienna, who had this power, took the preliminary decision until the final decision of the government of not putting Bertalanffy on the list of former Nazis. Consequently he was allowed to teach, and, among other things, made good use of the respite of five months for finishing his *Biologie für Mediziner* and his *Biologie und Medizin* until October. Nevertheless, he was again informed of his suspension in late January 1946. And he was notified that, from February until the clarification of his case, he would only receive the treatment provided for all those in his situation—namely the low sum of 150 schillings per month, equivalent to less than a quarter of his previous salary.

7.2.2 The Report of the Committee

The special committee charged with the examination of his case met on February 4, 1946. It produced a relatively uncritical report with regard to the facts.

The reason for that was that it accepted Bertalanffy's version of justifying his membership in the party. The committee did not have documents that refuted his version without ambiguity, in particular articles published by Bertalanffy in Germany, which, such as that of 1941, explicitly point to the congruence of his "organismic" philosophy and Nazi ideology—even if this claim is essentially an expression of opportunism.

By contrast, the committee had letters in which Bertalanffy accused Austria of having impeded the advancement of his career between 1933 and 1937 because of his sympathy for Nazism. It considered, in view of these letters, that he had lied to the Nazi administration in order to assure his livelihood and that this was "understandable". It noted that he had trampled on truth in an uncommonly unscrupulous manner in order to achieve his goals and that his behavior in this respect was intolerable, but accorded to him that he could not have predicted how far he had to go in this direction. The more serious, according to the commission, were his shameless lies related to the accusations against Austria.

Nevertheless, it concluded in a very positive manner that Bertalanffy must not be excluded from a reintegration into the university: it underlined in this sense that he had not made any concession to Nazi ideology of a non-scientific character in his domain; that on the contrary he had shown great strength of character and had served science by constantly trying to preserve the independence of biology. The committee charged the competent authorities, i.e., the Ministry of Education, with judging the pertinence of possible sanctions to be taken against him (Hofer, 1996: 34-36).

7.2.3 Purgatory and the First Desire to Emigrate

While waiting for this judgement, Bertalanffy requested in early March 1946 that he be paid his full salary, asking also for financial assistance to remedy the loss of his goods and his catastrophic situation. For this purpose he based himself on the preliminary decision of the mayor of Vienna and gave his efforts to restore the Department of Zoology as a reason. He reiterated his request on March 21, after the first had remained unanswered, demanding again that he be retroactively paid his normal salary from May to October 1945, which he had received only partially. But it was again in vain. He had to wait until the minister gave a ruling on his case, which only happened one and a half years later (Hofer, 1996: 36-37).

All in all, from January 1946 to December 1947, Bertalanffy was thus supposed to live on the salary that was given to him, the largest part of which was only enough to pay his and his wife's considerable tobacco and coffee consumption… During this period the Bertalanffy family survived essentially thanks to the generosity of English and American friends such as Woodger, who regularly sent food parcels. And an extra payment came from the articles of popular science which Bertalanffy and his son published in a Viennese weekly newspaper (*Die Woche*) (M. von Bertalanffy, 1973: 45; Davidson, 1983: 58).

Under these conditions Bertalanffy planned to emigrate (M. von Bertalanffy, 1973: 46). He reactivated his international contacts for this purpose from January 1946 onwards by writing in particular to the Rockefeller Foundation. But the research director in medicine of this foundation, Gerald H. Pomerat, asked him to wait until his visit to Vienna at the end of the year (Brauckmann, 1997: 10).

Later his motivation for emigrating increased all the more as he was not the only one to suffer from the consequences of his past actions. He knew the difficulties that awaited him professionally all the better, as his friend Gessner wrote to him on December 6, 1946, in order to describe his own, perhaps even worse, difficulties. Gessner told him how in Munich, where he taught, all scholars who had accepted to collaborate with the Nazis (like himself) experienced considerable difficulties in reintegrating into the university—to the extent that in his field, botany, no courses were offered anymore, and that for his part he could not expect to obtain a post before long, being reduced to give private seminars at his home.[3]

7.3 Bertalanffy's Work during a Period of Uncertainty
7.3.1 His Actual Publications, Pending and in Progress

Bertalanffy published very little between 1946 and 1947. In 1946 he published his first article in the journal *Nature* (a short study on the metabolism of planarians) and a short history of botany and zoology in Austria as well as a discussion concerning the relation between science and art in Austrian journals.

In the same year he also managed to publish his *Biologie und Medizin*; his analogous efforts as regards *Biologie für Mediziner* were, by contrast, in vain. Besides economical problems, a major reason for this was that the committee of "denazification" had not decided upon his fate. He shared in this respect again the worries of Gessner, three books of whom were still waiting to be published for the same reason[4] in early 1948. In fact, *Biologie für Mediziner* was never published at all.

---

3. Letter from F. Gessner to L. von Bertalanffy (06/12/1946), BCSSS archive.
4. Letter from F. Gessner to L. von Bertalanffy (14/01/1948), BCSSS archive.

A similar but in the end more happy fate awaited *Auf den Pfaden des Lebens*,[5] a popularization of his ideas that he wrote between 1947 and 1948 and that was only published in 1951. Bertalanffy spent most of his time in 1947 not with this work, however, but with the writing of a book which was to constitute one of his major works: *Das biologische Weltbild* (English title: *Problems of Life: An Evaluation of Modern Biological Thought*).

7.3.2 The Alpbach Meeting: The Rebirth of the Project of a General Systemology and the Beginnings of a Philosophical Anthropology

Bertalanffy also responded favorably to an invitation to the third "International College Weeks" ["Internationale Hochschulwochen", later: "Europäisches Forum Alpbach"] in Alpbach in 1947. This annual meeting, founded right after the war, aimed at providing an intellectual forum for "Young Europe" and at thereby promoting international exchanges without ideological boundaries.[6] It was organized in thematic working groups and Bertalanffy was invited, and accepted, to direct the "circle of biological work". The meeting took place from August 27 to September 6, 1947. Bertalanffy divided the discussion programme of his working group into three main parts. Even though the first, devoted to the "organismic" perspective and to its applications in all areas of biology, was only a general resumption of his work prior to 1938, the other two reflected the beginning of a transformation of Bertalanffy's centre of interest.

The second saw thus the emergence of a new topic in his thinking which became central later: the "specific position" of man in nature. It marked the beginning of a project, manifestly nurtured by Cassirer, which

---

5. Letter from L. von Bertalanffy to Schönfeld (26/04/1949), BCSSS archive.
6. Letter from W. Blaser (Generalsekretariat Österreichisches College) to L. von Bertalanffy (29/04/1948), BCSSS archive.

he described as "anthropological philosophy". While criticizing biologism and being careful not to yield to it, Bertalanffy outlined, in the extension of his reflections on the specificity of human morphogenesis introduced in 1937, the biological foundations of human behavior. They are determined by evolution (a relatively low degree of specialization of certain organs, the specificity of the hormonal system, an increase of the brain size), development (a relative delay of the development resulting from the hormonal specificity which implies a long period of youth conducive to mental development), and particular neurological characteristics. These features, which make of man the only animal without organic and instinctual adaptation to a specific environment, offer him in return the ability to overcome what may appear as a set of biological defects, to make "a virtue out of his biological helplessness" (L. von Bertalanffy, 1956a: 35): that of creating his own environment—culture. Man is a "symbolic animal": he gets his specificity from his power to substitute a world of "things" with a universe of symbols. And it is precisely because he does not experiment with "things" anymore but with their symbols that culture imposes itself on nature and that History supplants phylogenetic evolution. Ordered into a system by a "grammar" that codifies their relationships, the symbols engender universes with their proper logic. This is on the one hand the source of the fertility of these universes, provided that the symbols and their "grammar" are well chosen: the symbolic systems become "more intelligent" than their creators, in the sense that the results of the symbolic operations can correspond to actual events, of whose existence man is unaware—Bertalanffy took up here, in particular, the conceptions of the physicist Heinrich Hertz (1857-1894). But this immanent logic of symbolic systems is also the source of a permanent danger: it may produce conflicts between symbolic universes and biological instincts, or between the symbolic universes themselves, leading in the first case to individual neuroses and in the second to ideological and political conflicts. A bivalent reflection, marked by the recent political tragedies, was initiated in Alpbach and was to allow Bertalanffy to rapidly attract the attention of American sociologists, psychologist and psychiatrics who developed convergent conceptions. That is to say on the one hand the rooting of psychological and social problems of humans in biological factors connected to their unique evolution—a perspective that led him for example to return to the theses of Trigant Burrow (1875-1950);[7] and on the other hand, the insistence on the irreducibility of the human universe to its biological determinations, and on the particular dignity that derives from it.

What began to take shape underlying his ideas was also an awareness of what he recognized in private to be the "dialektische Tragik des Ganzheitsbegriffes" (the dialectical tragedy of the concept of wholeness), while embodying the affirmation of individuality in its own logic in relation to the environment, the fullness and unity of human experience—the romantic *Hen Kai Pan* ("One and All") and the ideal of an education of man in his entirety—this concept was always threatened by an "uncritical transfer to the human community that leads to the extermination of the individual and to a totalitarian State".[8] Such reflection is certainly not an expression of a break with his previous philosophy: it rather renews the form of his critical analyses on the "mystique of wholeness" [Ganzheitsmystik], which were constitutive for his work and explicitly postulated as such in his writings until 1940. But, and this is peculiar to the "tragedy", this new awareness did not prevent the destiny of this

---

7. Correspondence between L. von Bertalanffy and T. de Burrow (22/10/1948 – 25/03/1949); letter from H. Syz to L. von Bertalanffy (07/09/1950), BCSSS archive.
8. Letter from L. von Bertalanffy to K. Buchwald (04/04/1950); letter from L. von Bertalanffy to B. Lustig-Olthuis (23/06/1952), BCSSS archive. Translated from the German by ES.

concept to be fulfilled with Bertalanffy; and one saw him correspond with Keiter still in 1950, at a moment when Keiter converted his "(hygienic and racist) biology of culture" into a systemic theory of history.[9]

At the Alpbach meeting, the third part of Bertalanffy's work was a lecture he held on September 4, entitled "Einheit der Wissenschaft und Prinzipien einer allgemeinen Systemlehre" (The Unity of Science and Principles of a General Systemology) and followed by a debate. This lecture was essentially a reprise of his article on the subject that he had not been able to publish in 1945. It thus constituted his first public talk on general systemology since 1937—and produced in 1948, with another much shorter one (in the journal *Europäische Rundschau*), his first article published on the subject. This lecture, and more generally the Alpach meeting, provided Bertalanffy with the possibility of fruitful discussions about the idea of a general systemology with Austrian scholars coming from different academic fields, in particular with the physicist Arthur March.

7.4 The Results of the Procedure of "Denazification"

The procedure of "denazification" was also running its course during this period. But it reached its end only slowly because it got lost in the maze of administration. A first report of the Ministry of Education dated October 22, 1946, affirmed that nothing had been established that would allow the conclusion that Bertalanffy was guilty. But still in early 1947 the rector demanded from the president of the committee of "denazification" responsible for the case to send him all its elements in order that a definitive decision could be made. Again on June 13 Bertalanffy submitted a request with

---

9. Correspondence between L. von Bertalanffy and F. Keiter (19/01/1949 – 04/04/1950), BCSSS archive.

Chapter 7: His Last Years in Vienna

the federal presidents, but did not receive a response for the time being. The final decision concerning him was only made on October 25, 1947. It stipulated that he was exempted from prosecution and that this decision was to be applied retroactively from February 18, 1947 onwards, giving him the corresponding rights (Hofer, 1996: 38).

Bertalanffy thus returned to the university in all normality in early December 1947, but only as a lecturer without civil servant status [*Privatdozent*] (Brauckmann, 1997: 9)—i.e., at the age of 46 precisely with the status he had when he had been between 33 and 37 years old. One of his first lectures clearly indicated the direction of his thinking during that period: entitled "The Meaning and Unity of Science", it was a talk about the principle ideas of his general systemology. Very understandably however, Bertalanffy soon began to take steps to obtain a promotion to the rank of associate professor, which would allow him to regain his previous status. In June 1948 these steps led to a decision by the university that was in principle positive (Brauckmann, 1997: 9), which the Ministry, however, had yet to endorse.

CHAPTER 8
Preliminaries of a Definitive Emigration (1948-1949)

8.1 An Invitation to London

Despite this reintegration, Bertalanffy could not bear the humiliation that was imposed on him with the examination in the committee and the deterioration in his status, nor the relations within the university and the living conditions in Vienna.[1] But at the same time other horizons appeared, which maintained his only true hope in more than two years: finding a post abroad. The reactivation of his contacts in the United States seemed to bear fruit for a while. In fact he received a telegram dated October 28, 1947, with the proposition to work in Cleveland as director of research in the biological division of the clinic run by a certain Mr. Quiring. As Bertalanffy saw no possibility of obtaining a visa in Vienna by himself, he wrote to the Rockefeller Foundation in order that a work permit might be sent to him to the United States. Unfortunately for him, there was no response to his letter.

A real opportunity came a few months later from his friend Woodger, who knew his difficulties and who, on April 10, 1948, sent him an invitation to come and work at his side for one academic year in the department of biology that he was in charge of at the *Middlesex Hospital* in London in order to help, in particular, to develop his "axiomatic biology". Woodger did everything in his

---

[1]. Letter from L. von Bertalanffy to W.W. Nowinski (30/01/1950); letter from L. von Bertalanffy to A. Mittasch (03/02/1949), BCSSS archive.

power to arrange for his visit the status of visiting professor and obtained the official authorization of the financial committee of his hospital on May 21, of which he informed Bertalanffy on May 31.[2] The latter straight away sent a request for temporary leave of absence (with salary) of six months to the administration to which he belonged at the University of Vienna, a request that was quickly accepted, with his leave starting officially on August 24 (Hofer, 1996: 39).

8.2 An Intermediate Journey to Switzerland and a Scholarship for Canada. *Das Biologische Weltbild* and its Reception
8.2.1 Bertalanffy's Activities in Switzerland

Before going to England, Bertalanffy had to complete his courses of the second semester of 1947/1948 in Vienna. Then he left with his wife and his son for Switzerland. He was in fact invited in July by the economist Walter A. Jöhr (1910-1987), who proposed to him to completely take care of his journey and that of his family, to permit him to finish *Das Biologische Weltbild* in good conditions, and also to publish this book.

No sooner had Bertalanffy arrived on the scene than he learned that his contacts in America had finally borne fruit: a scholarship from the Davis Foundation of five thousand dollars for coming to work in Canada was granted to him on July 17, 1948, together with the payment of the journey. The Foundation had also contacted the dean of the University of Toronto in order to secure a post for him.[3] Thus the hope of a definitive emigration

---

2. Letters from J.H. Woodger to L. von Bertalanffy (10/04/1948 and 31/05/1948), BCSSS archive.
3. Letter from B. Hayes (Lady Davis Foundation) to L. von Bertalanffy (17/07/1948), BCSSS archive. Brauckmann,

Chapter 8: Preliminaries of a Definitive Emigration

finally found its full satisfaction.

The family was accommodated in a hospice in Bern and stayed there for six weeks. During this period Bertalanffy gave several lectures in Basle. He also intended to publish a revised edition of his book on Nicholas of Cusa from 1928. He took steps in this direction with the publisher *Atlantis* in Zurich. The latter received his request favourably, while refusing to publish an essay on Goethe's philosophy of nature that Bertalanffy submitted. Paradoxically, the book on Cusa would never be published by *Atlantis*, while the essay on Goethe would: as to the first, Bertalanffy considered the financial proposition (800 DM) too low,[4] and he revised the second, which would be published in August 1949. Anxious to see the encyclopaedic work, whose publication he initiated in 1942, succeed, Bertalanffy also suggested during this stay that Gessner should follow him to ensure the resumption of the publication of the *Handbuch der Biologie*, which his friend accepted.[5] The editor's agreement on this resumption[6] arrived in October 1949. Finally Bertalanffy began to work at a revision of his *Theoretische Biologie* in Berne. And above all, as settled with Jöhr, he completed his last corrections of *Das Biologische Weltbild*.[7]

1997: 10.

4. Letter from *Atlantis Verlag* to L. von Bertalanffy (29/10/1948) and from L. von Bertalanffy to *Atlantis Verlag* (08/11/1948), BCSSS archive.

5. Letter from F. Gessner to L. von Bertalanffy (08/09/1949), BCSSS archive.

6. Letter from L. von Bertalanffy to F. Gessner (28/10/1949), BCSSS archive.

7. Letter from L. von Bertalanffy to Schönfeld (26/04/49), BCSSS archive. M. von Bertalanffy, 1973: 46; Hofer, 1996: 39; Brauckmann, 1997: 10.

8.2.2 "The Biological Conception of the World"

This last work was the synthesis of Bertalanffy's reflections during the past twenty years, a systematization of topics he had previously developed in his various writings. Even though this synthesis, as regards his "organismic" philosophy as such, did not bring in anything essential in relation to *Das Gefüge des Lebens* (1937), it constituted a successful rendering of his ideas into a language that tends to substitute Goethean verses with the technicality of proper scientific discourse.

The originality of this book compared to his earlier writings lies in the fact that Bertalanffy, in the spirit of the project of a general systemology, which constituted henceforth the heart of his intellectual concerns, did not only argue for the necessity of an "organismic" point of view in all domains of biology—in particular by demonstrating the relevance and the fertility of the concepts of the "open system" and of "hierarchical order". He also tried to establish the convergent evolution of all natural, social, and human sciences as well as of philosophy towards an epistemology centred on the concepts of the system and of dynamic organization. The logic of *Das Biologische Weltbild* is the embodiment of Bertalanffy's fundamental idea and the culmination of a journey initiated in his doctoral thesis of 1926: surpassing organicism in the direction of systemism. His book accordingly concludes with an exposition of the main principles of his general systemology.

Even if this dimension is not explicitly stated in this work, Bertalanffy also considered it, in the extension of his reflections initiated in Alpbach, as the inauguration of a "philosophical anthropology based on the concepts of wholeness and the symbol".[8] *Das Biologische Weltbild*

---

8. Letter from L. von Bertalanffy to K. Buchwald (04/04/1950), BCSSS archive.

was in fact announced as a first volume and appeared in early 1949 with not only the announcement of a second volume, but also that of its detailed plan—as this second volume was precisely devoted to developing this philosophical anthropology, whose main concerns had been presented in Alpbach. Nevertheless, the conditions for an emigration would thwart this project and Bertalanffy would be confined to publish the already finished parts in the form of articles.[9]

8.2.3 The Reception of *Das Biologische Weltbild*

The "first volume" received only limited response in its original (German) version: it was welcomed with enthusiasm by scientists,[10] doctors[11] and philosophers[12] who had already had affinities with Bertalanffy's "organismic" thought and some of whom considered him even as "the greatest biologist alive"[13] or as "the Einstein of biology".[14] He also aroused the interest of several economists: that

---

9. Letter from L. von Bertalanffy to A. Auersperg von (15/05/1950), BCSSS archive.
10. Letter from A. Mittasch to L. von Bertalanffy (10/06/1949); letter from F. Gessner to L. von Bertalanffy (08/09/1949 and 01/10/1949); letter from B. Klatt to L. von Bertalanffy (25/08/1949); letter from A. Stäger to L. von Bertalanffy (21/12/1950); letter from L.H. Taylor to L. von Bertalanffy (12/09/1952), BCSSS archive.
11. Letter from A. Auersperg von to L. von Bertalanffy (01/09/1950); letter from F. Erbslöh to L. von Bertalanffy (16/01/1952); letter from W.O. Gross to L. von Bertalanffy (18/08/1952), BCSSS archive.
12. Letter from E. Rothaker to L. von Bertalanffy (25/07/1949); letter from F. Keiter to L. von Bertalanffy (20/06/1949), BCSSS archive.
13. Letter from G.C. Hirsch to L. von Bertalanffy (05/06/1952), BCSSS archive.
14. Letter from F. Gessner to L. von Bertalanffy (23/04/1950), BCSSS archive.

of Jöhr of course, particularly with regard to the idea of a general systemology, and that of the eminent Friedrich A. von Hayek (1899-1992), who emigrated from Vienna to London and with whom Bertalannfy maintained correspondence—so that Hayek, who found in the latter a strong convergence with his own philosophy of knowledge, aroused his attention about an essay on the subject in 1950 of which he had only drafted an outline.[15]

But the reception of *Das Biologische Weltbild* suffered from an ideological context that was much less favorable, from a shift in the centre of gravity of scientific activity in America and the arrival of molecular biology. In 1949 the physiologist Erwin Bünning (1906-1990) published a harsh critique in one of the major German scientific journals (*Naturwissenschaften*), which denounced among others Bertalanffy's claim to be the father of the "organismic" conception. And in turn, the publication of a very favorable evaluation submitted by Gessner to another important journal (*Zeitschrift für Naturforschung*) was refused, on the initiative of Max Hartmann,[16] another "enemy" of Bertalanffy. In fact, even though Bertalanffy was still known in Germany and Austria during the years after his departure, he was far from being popular,[17] with the exception of some circles that kept supporting the "organismic"[18] theses and those that, like Jöhr and Hayek, saw in his thought interesting possibilities of interaction between biology and social sciences that go beyond empty and dangerous analogies.

The publication of his article on general systemology, which could not appear in 1945 and which he

---

15. Letters from F.A. Hayek to L. von Bertalanffy (14/05/1949 and 15/04/1950), BCSSS archive.
16. Letter from F. Gessner to L. von Bertalanffy (16/11/1949), BCSSS archive.
17. Letter from L. von Bertalanffy to F. Gessner (12/11/1951), BCSSS archive.
18. Letter from F. Gessner to L. von Bertalanffy (01/07/1952), BCSSS archive.

published (with some significant changes) in the journal *Biologia Generalis* (which he directed) in 1949, also suffered the same fate. Not without bitterness he would note in 1952 that few Germans and Austrians seemed to be interested in it, although it was met with "great interest" in America.[19]

8.3 The Journey to Great Britain

Bertalanffy declined a second invitation to the annual Alpbach meeting[20] that took place from August 21 to September 9, 1948, in order to go to London with his wife on August 23. As their son Felix had not received a work permit in England, he was unable to obtain a visa: while waiting for being able to join his parents, he stayed in Zurich, where he worked as the assistant of a surgeon.

8.3.1 Another Ministerial Rejection of the Request for Promotion at the University of Vienna and the Renewal of the Leave of Absence at this University

One month later Bertalanffy had news from Vienna that confirmed him in his decision to emigrate: his nomination as an "associate professor", endorsed by the university in June, was finally refused by the Ministry of Education on November 10, 1948 (Brauckmann, 1997: 9-10).

But he was still attached to the University of Vienna with his status as a lecturer on leave of absence and he did not hesitate to ask for an extension of one year of this status for the period from February 28, 1949, to February 28, 1950. His request was accepted, with the

---

19. Letter from L. von Bertalanffy to F. Gessner (19/09/1951), BCSSS archive.
20. Letter from W. Blaser (Generalsekretariat Österreichisches College) to L. von Bertalanffy (29/04/1948), BCSSS archive.

significant restriction that his salary was reduced (Hofer, 1996: 39). He would later repeat the same request every six months for the following six years, a request which was always satisfied (but with cancellation of his salary). But, warned in June 1956 by the director of the Institute of Zoology of the illegality of this situation (a maximum of two years was actually allowed), the dean of the Faculty of Philosophy intervened on February 4, 1957, asked Bertalanffy whether he intended to come back to Vienna in order to return to his position as a lecturer, and threatened him with expulsion from the university if he did not. Bertalanffy, replying to this formal notice on March 24, 1957, was offended to see that one wanted to remove his name in Vienna when, according to him, foreign universities took pride in having a person whose works were recognized worldwide on their list of professors. And, accepting a return only under the condition of having the title of professor, he preferred to resign, as the intention of the University of Vienna did obviously not go into this direction (Hofer, 1996: 41; Brauckmann, 1997: 10, 233-244).

8.3.2 The Residence Permit in Canada

Luckily for Bertalanffy, the last possible obstacle to his emigration was removed almost simultaneously with the rejection of the Austrian Ministry of Education to confirm his promotion: the permission to live in Canada was issued on November 30, 1948, after more than four months of administrative measures that had tried his patience. This permission was also issued to his wife; but their son still had to wait a few more months for it.[21]

---

21. Letter from the *Canada Department of Mines and Resources* to L. von Bertalanffy (30/11/1948); letter from B. Hayes (Lady Davis Foundation) to L. von Bertalanffy (27/09/1948), BCSSS archive.

Around the same time Bertalanffy equally learned that he would ultimately not be assigned to the University of Toronto, but to the Department of Zoology at McGill University, Montreal, as a visiting professor (Hofer, 1996: 40).

8.3.3 Bertalanffy's Activities in Great Britain: The Publicity of the Project of General Systemology in the Anglo-Saxon World and Its Reception

As to his stay in Great Britain itself, it lasted five and a half months. Bertalanffy seemed, however, not to be limited to the British capital, but was, for instance, invited to Edinburgh on November 17, 1948, in order to give several lectures—Waddington (who directed the Institute of Genetics there) was delighted to welcome him.[22]

In London Bertalanffy enthusiastically discussed his views on the philosophy of biology with Woodger. But he also talked a lot about his general systemology and was happy that it was met with great interest, being described by some of his hosts as "fantastic" and "of outstanding importance".[23] In particular he gave a lecture about "the concept of the system in physics and biology" on January 17, 1949, at the second meeting of the *Philosophy of Science Group*, which had been formed the previous year within the framework of the *British Society for the History of Science*. Bertalanffy had the honour to have the illustrious philosopher Bertrand Russell (1872-1970) among his audience (Davidson, 1983: 59; Brauckmann, 1997: 10), of whom Woodger was a spiritual heir. This lecture was published in 1950 in the journal *Science*. Another article on his general systemology, which

---

22. Letter from F. Gross to L. von Bertalanffy (17/11/1948), BCSSS archive.
23. Letter from L. von Bertalanffy to Schönfeld (26/04/1949), BCSSS archive.

largely took up the ideas published in German in 1949 and aimed at a systematization of this lecture,[24] was also published in 1950, in the first issue of the *British Journal for the Philosophy of Science*, edited by the Philosophy of Science Group.

These two articles soon made Bertalanffy's views known throughout the world, especially in the Anglo-Saxon world. They aroused great interest and many favorable opinions among biologists,[25] chemists,[26] physicists,[27] psychologists,[28] philosophers[29] and even geographers[30]—across the Atlantic as well as in Europe. Fearing that the second article might remain unnoticed in the United States, Bertalanffy tried to have it published simultaneously in the American journal *Philosophy of Science*, but suffered a rejection of the editor Charles W. Churchmann (1913-), who considered it an unnecessary duplication.[31]

---

24. Letter from L. von Bertalanffy to F.A. Hayek von (02/05/1950), BCSSS archive.
25. Letter from G. Hardin to L. von Bertalanffy (19/11/1952), BCSSS archive.
26. Letter from K.E. Shuler to L. von Bertalanffy (19/04/1951), BCSSS archive.
27. Letter from J.D. Trimmer to L. von Bertalanffy (09/01/1951), BCSSS archive.
28. Letter from D. Krech to L. von Bertalanffy (24/10/1950); letter from G.T. Thomas to L. von Bertalanffy (02/06/1951), BCSSS archive.
29. Letter from F.I.G. Rawlins to L. von Bertalanffy (20/03/1952); letter from B. Taylor to L. von Bertalanffy (01/11/1952), BCSSS archive.
30. Letters from E. Winkler to L. von Bertalanffy (09/05/1949 and 15/07/1949), BCSSS archive.
31. Letter from W. Churchmann to L. von Bertalanffy (09/05/1950); letter from L. von Bertalanffy to F.A. Hayek (02/05/1950), BCSSS archive.

CHAPTER 9
Arrival in Canada: Montreal and Ottawa (1949-1954)

9.1 Six Months in Montreal

In early February 1949 the whole Bertalanffy family boarded a luxury liner in order to cross the Atlantic (M. von Bertalanffy, 1973: 46), and Bertalanffy took up his post at McGill University, Montreal, on February 15 (Brauckmann, 1997: 11). His son Felix enrolled to do a master's degree in histology[1] and afterwards stayed there to do his doctorate under the supervision of Charles P. Leblond (1910-).[2]

In Montreal Bertalanffy worked under "very satisfying" conditions, all the more so as he received financial aid in addition to his grant so that he could compensate the loss of his property in Vienna.[3] He had discussions concerning the application of the concepts of the open system and the steady state to the problem of the interaction between blood plasma and histological fluids with the biochemist D.L. Thompson, who was also the dean of the faculty.[4] He also had the opportunity to meet the German mathematician Hans Zassenhaus (1912-1991), an algebraist who had also just arrived there and who was

---

1. Letter from L. von Bertalanffy to M. von Auersperg (29/01/1950), BCSSS archive.
2. Letter from L. von Bertalanffy to A. von Auersperg (15/05/1952), BCSSS archive.
3. Letter from L. von Bertalanffy to Schönfeld (26/04/1949); letter from L. von Bertalanffy to F. Werner (19/04/1949), BCSSS archive.
4. Letters from L. von Bertalanffy to D.L. Thompson (08/11/1950 and 13/11/1950), BCSSS archive.

9.2 Bertalanffy's Preliminary Negotiations for His Transfer to Ottawa

But Bertalanffy only stayed in Montreal for six months. He was presented with an opportunity at the University of Ottawa, where the function of the director of the Department of Biological Research in the newly founded Faculty of Medicine was vacant and offered to him. In agreement with the president of McGill University he got in contact with the administrative director of this faculty, the Reverend A.L.M. Danis, on May 7, 1949, in order to arrange the conditions of his transfer.

The main reason why Bertalanffy was interested in this post was that it provided him with a guaranteed security in the long term. He demanded a five-year contract and set the conditions for his salary.[6] At the end of May he met Danis in Montreal, who accepted these conditions as well as one other condition, namely that Bertalanffy would become the director of the Department of Biology, since Le Bel, who then held this position, would soon retire. Bertalanffy, who claimed to have received other proposals in Canada and the United States, only accepted the post when assured that he would "not be subordinated to anybody", considering that another position was "out of question" for him given his "international reputation".[7]

Satisfied by the (oral) agreement with Danis, Bertalanffy submitted a detailed plan for the reorganization

---

5. Letter from B. Klatt to L. von Bertalanffy (25/08/1949), BCSSS archive.
6. Letter from L. von Bertalanffy to A.L.M. Danis (19/05/1949), Archive of the University of Ottawa.
7. Letter from L. von Bertalanffy to A.L.M. Danis (06/05/1950), Archive of the University of Ottawa.

Chapter 9: Arrival in Canada: Montreal and Ottawa

of the Department of Biology on June 14, including the calculation of the budget necessary for this purpose. Bertalanffy wanted to orient the activities of the department on the one hand radically towards experimental research and the corresponding teaching so as to be able to pursue his work on the physiology of growth; and on the other hand, to reproduce what he had already constructed in Vienna during the war, namely the creation of a unique department of biology that integrated zoology, botany, and medicine.[8] He also suggested that the English translation of *Das biologische Weltbild* should be published by the University of Ottawa, and that the journal *Biologia Generalis* and the *Handbuch der Biologie* should be coedited by this university. His ambition was to turn the Department of Biology into one of the leading Canadian scientific institutions. Except for his request that his son be admitted to do his doctorate at the Faculty of Medicine in Ottawa,[9] all his requests were acceded to and Bertalanffy accepted his nomination on August 30.[10]

9.3 Bertalanffy's Activities in Ottawa and the Reception of his Work

Bertalanffy and his wife came to Ottawa and settled there on September 6. And, in order to ensure a comfortable income, Bertalanffy employed his own wife as an assistant.[11]

---

8. Letter from L. von Bertalanffy to A.L.M. Danis (14/06/1949). Copy of a plan for the Department of Biology and of Medical Biology (30/06/1950), Archive of the University of Ottawa.
9. Letter from L. von Bertalanffy to A.L.M. Danis (14/06/1949), Archive of the University of Ottawa.
10. Letter from L. von Bertalanffy to A.L.M. Danis (30/08/1949), Archive of the University of Ottawa. Letter from L. von Bertalanffy to M. von Auersperg (29/01/1950), BCSSS archive.
11. Davidson, 1983 : 60-61. Letter from L. von Bertalanffy to

He provided courses in biology for medical students at this university and performed the hard work of establishing and organizing the Department of Biological Research, for which he received substantial subsidies. The scientific work throughout the almost five years that Bertalanffy spent in Ottawa was oriented towards five main directions.

9.3.1 The Development of the Theory of Open Systems

He developed first of all a genuine biophysics of open systems. This development led to the publication of his book *Biophysik des Fliessgleichgewichts* (Biophysics of the Steady State) in 1953, which was a consolidation of his founding article of 1940 in the light of recent work concerning the thermodynamics of irreversible processes—in particular that of Ilya Prigogine (1917-2003) and Kenneth G. Denbigh,[12] who had referred to him very early, and that of Sybren R. de Groot. In this work his Heraclitean and Goethean vision of a living nature whose permanence is only an order based on a perpetual flow of matter and energy is combined with the principles of the kinetics and thermodynamics of open chemical systems, from which the fundamental properties and certain quantitative laws of living systems are deduced. Bertalanffy thus tries to show that any biological phenomenon, including evolution, does not violate the laws of physics, provided that the latter is supplemented by the theory of open systems.

The correspondence with the Austrian chemist Anton Skrabal (1877-1957), which began in July 1949 and which allowed him to get to know better the work

---

M. von Auersperg (29/01/1950), BCSSS archive.
12. Letters from K.G. Denbigh to L. von Bertalanffy (05/12/1951 and (28/05/1952); letter from L. von Bertalanffy to F. Gessner (19/05/1951), BCSSS archive.

of the latter concerning the kinetics of simultaneous chemical reactions, also contributed significantly to his reflections on the concept of the open system, which was the subject of detailed exchanges between them. It also strengthened him in his belief in the relevance of his general systemology,[13] insofar as Skrabal, in his field, used a formalism and expounded principles that were transferable to other domains and that could thus be perfectly integrated into his project.

Bertalanffy equally tried to arouse the interest of the eminent physico-mathematician Hermann Weyl (1885-1955) by writing to him in early 1952 in order to introduce his work to him and to ask his opinion regarding his work on the kinetics and thermodynamics of open systems. Nevertheless, Weyl was especially interested in his general systemology, while pointing out to him its fragmentary nature. And he introduced to him in return his book *Symmetry*, which filled Bertalanffy with enthusiasm about the "unification of mathematics, philosophy and aesthetics" that was realized there.[14] But there was no continuation of this correspondence.

If *Biophysik des Fliessgleichgewichts* had hardly any success in German-speaking countries ("the physicist believes that biophysics does not concern him and the biologist is frightened off by the formulas")[15] and was *a fortiori* completely unknown in America, this was not true for an article Bertalanffy published in the journal *Science* in 1950, where he expounded, by already taking into account the thermodynamics of irreversible phenomena, the outline of this theory of open systems and its purpose.

---

13. Correspondence between L. von Bertalanffy and A. Skrabal (30/07/1949 – 24/06/1952), in particular letters on 28/08/1949 and 21/09/1949, BCSSS archive.
14. Correspondence between L. von Bertalanffy and H. Weyl (21/03/1952 – 19/12/1952), BCSSS archive.
15. Letter from L. von Bertalanffy to W. Westphal (18/11/1955), BCSSS archive. Translated from the German by ES.

The Dialectical Tragedy of the Concept of Wholeness

This article, as well as a review in *Nature* in 1949 about the work of Prigogine, made him widely known on the New Continent and even in Japan.[16] There were numerous physicists,[17] chemists,[18] biologists,[19] engineers,[20] but also psychologists, sociologists[21] and philosophers who assessed this theory very positively and, for their part, helped Bertalanffy progress in the development of his ideas. Two and a half months after its publication, almost three hundred letters on the subject, almost all of them enthusiastic, reached Bertalanffy,[22] while he contributed through his responses to the introduction of the work of Prigogine, Denbigh and de Groot in America.

In particular, the mathematician Carl Lienau contacted Bertalanffy after having read with great interest his article in *Science*. He acquainted him with the mathematical theory of organization (based on the theory of groups) that he had published in 1947, a theory that Bertalanffy considered "exciting" and to which he would henceforth refer regularly. They maintained their correspondence for a certain time and Lienau suggested to him to contribute, instead of the deceased Lotka, to a collective book on the

---

16. Letter from H.K. Kihara to L. von Bertalanffy (08/08/1952), BCSSS archive.
17. Letter from S.T. Bornemisza to L. von Bertalanffy (31/03/1950), BCSSS archive.
18. Letters from K.G. Denbigh to L. von Bertalanffy (05/12/1951 and 28/05/1952), BCSSS archive.
19. Letter from the *Canada Department of Agriculture* to L. von Bertalanffy (18/04/1950); letter from E.H. Battley to L. von Bertalanffy (11/11/1951); letter from M. Graham to L. von Bertalanffy (30/05/1949); letter from H.F. Rosene to L. von Bertalanffy (14/11/1951); letter from F. Warburton to L. von Bertalanffy (16/06/1952), BCSSS archive.
20. Letter from E.W. Leaver to L. von Bertalanffy (25/10/1951), BCSSS archive.
21. Correspondence between L. von Bertalanffy and H.O. Englemann (04/04/1950 – 18/07/1951), BCSSS archive.
22. Letter from L. von Bertalanffy to F. Keiter (04/04/1950), BCSSS archive.

Chapter 9: Arrival in Canada: Montreal and Ottawa

"theory of organization" that he had planned with the latter during the war. Bertalanffy accepted this proposal but there was no follow-up to this agreement.[23]

The American philosopher Arthur F. Bentley (1870-1957) took the initiative to start a fruitful correspondence with Bertalanffy, which took place between 1959 and 1952. He wrote to him on August 20, 1950, in order to communicate to him his enthusiasm about his theory of open systems, which, according to him, was from the point of view of the theory of knowledge made from the same material as the views of Percy W. Bridgman (1882-1961), of Bohr, and especially of John Dewey (1859-1952) and of himself. Bentley believed, and Bertalanffy fully shared this opinion, that the "transactional" point of view, which he promoted together with Dewey, abolished—just like Bertalanffy's "organismic" thought—the barrier between organism and environment by substituting it with the "Heraclitean flux". Bentley introduced his collaborative work with Dewey to Bertalanffy (among others *Knowing and the Known*), while communicating articles of the latter to Dewey, who was very interested in his general systemology.[24]

9.3.2 The Beginnings of Bertalanffy's Entry into the Fields of American Psychology and Psychiatry

Essential for the development of Bertalanffy's focus of interest was his impact in the field of psychology, the "tremendous interest" and the "excitement" that his theory of open systems and his general systemology aroused among those who tried to establish a psychology

---

23. Correspondence between L. von Bertalanffy and C. Lienau (24/01/1950 – 16/11/1951), in particular 24/11/1950 and 29/08/1950, BCSSS archive.
24. Correspondence between L. von Bertalanffy and A.H. Bentley (20/08/1950 – 25/11/1952), BCSSS archive.

away from behaviorism and who were for that purpose in search of new theoretical models.[25] This is what brought him particularly into contact with David Krech (1909-1977), who developed a theory of personality based on the concept of the open system in 1950 by referring explicitly to him. Bertalanffy, who considered this theory "fascinating" and found it "in complete agreement" with his conceptions, discovered then with enthusiasm that American psychology opened up to him,[26] while the German and Austrian counterparts had, with some exceptions such as Thumb, hardly manifested such interest. Krech then invited Bertalanffy to speak at a conference on the "models of personality", which he organized in Harvard in 1951.[27]

Bertalanffy gave a talk at this conference on the "theoretical models in biology and psychology", which would be followed by an article that signalled his entry into the world of American psychology. In this article he supplements his reflections on the concept of the model and the status of his general systemology as a theory of systemic modelization. But he tries above all to demonstrate the relevance of his "organismic" principles for psychology: not only that of the "open system", but also that of "hierarchization"—in particular of "progressive mechanization". He stipulates the creation of models in psychology that, in his view, need to be "essentially dynamic, although including structural order", "molar" although permitting "molecular" interpretations, and "formal" although allowing future "material" interpreta-

---

25. Letter from H. Cantril to L. von Bertalanffy (29/03/1950); letters from D. Krech to L. von Bertalanffy (11/10/1950 and 24/10/1950); letter from E. Brunswik to L. von Bertalanffy (14/03/1952), BCSSS archive.
26. Letter from L. von Bertalanffy to D. Krech (15/10/1950), BCSSS archive.
27. Letter from D. Krech to L. von Bertalanffy (24/10/1950); letter from L. von Bertalanffy to H.O. Englemann (18/07/1951), BCSSS archive.

tions (L. von Bertalanffy, 1951: 36). And he outlines an "organismic model of personality" whose further development would constitute a major part of his subsequent work.

What is important regarding this orientation towards psychology is also the fact that Bertalanffy was contacted by Abraham Maslow (1908-1970) in 1951, one of the future leaders of "humanistic psychology": he claimed to be an "old admirer" of his writings, was delighted to see him work on the New Continent and could not wait to meet him. And he introduced his work to him, which Bertalanffy regretted not to have known before, as it converged with his own.[28]

Bertalanffy's article in *Science* on open systems finally drew attention to him in the field of psychiatry, which also determined his future work: the psychoanalyst and psychiatrist Karl A. Menninger (1893-1990) wrote to him on June 4, 1951, that he was "very much interested" in his theory of open systems, which is according to him highly relevant for his own work, where the concepts of equilibrium and homeostasis played an important role. Bertalanffy found Menninger's work "stimulating" and "fascinating".[29]

9.3.3 The Pursuit of Experimental Research on Organic Growth

In Ottawa Bertalanffy also continued his research on the relationship between organic growth and metabolism. It is marked by a consistent series of experiments carried out with the assistance of the Polish physician and sur-

---

28. Letters from A. Maslow to L. von Bertalanffy (1951, before October) and from L. von Bertalanffy to A. Maslow (06/10/1951), BCSSS archive.
29. Letters from K.A. Menninger to L. von Bertalanffy (04/06/1951) and from L. von Bertalanffy to K.A. Menninger (18/07/1951), BCSSS archive.

geon W.J.P. Pyrozinski, whose arrival from Switzerland he had arranged even before he took up his post in Ottawa.[30] This work concerned the relative growth of organs, the experimental verification of its equations of global growth (especially in mammals) and the research of factors that control the respiration of tissues during growth. Twelve articles were published on these issues by his department throughout the period from 1950 to 1954.

These articles, preceded by a general statement that summarized his previous research published in the journal *Nature* in 1949, led Bertalanffy to quickly get in touch with a number of physiologists of the Anglo-Saxon world working on the same topics, who either had not known his work before[31] or were delighted to see it become established in America.[32] He also played an important role in another way in this domain, as he made American and Australian scholars acquainted with German-speaking scholars, such as his colleague and friend from Heidelberg Wilhelm Ludwig[33] (one of the leading

---

30. Letter from L. von Bertalanffy to A.L.M. Danis (08/06/1949), Archive of the University of Ottawa. Letter from L. von Bertalanffy to W.W. Nowinski (07/06/1950), BCSSS archive.

31. Correspondence between L. von Bertalanffy and R.R. Crandall & A.H. Smith (20/06/1952, 03/07/1952); correspondence between L. von Bertalanffy and M.F. Day (14/08/1951, 22/10/1951); letter from C. Ellenby to L. von Bertalanffy (30/01/1949); letter from L. von Bertalanffy to B. Grad (07/01/1951); correspondence between L. von Bertalanffy and M. Graham (08/04/1949 – 30/05/1949); letter from W. Guin to L. von Bertalanffy (10/07/1951); letter from J.P. Ide to L. von Bertalanffy (28/02/1952); letter from K. Schmidt-Nielsen to L. von Bertalanffy (18/06/1951); letter from F.W. Weymouth to L. von Bertalanffy (16/07/1951), BCSSS archive.

32. Letter from R.O. Earl to L. von Bertalanffy (05/01/1950), BCSSS archive.

33. Correspondence between L. von Bertalanffy and M.F. Day (14/08/1951, 22/10/1951), BCSSS archive.

experts on organic growth since the 1920s). For instance, he had one of Ludwig's students come to work with him in Ottawa at Ludwig's request (Krywienzcyk)[34] and suggested in late 1949 to publish the doctoral thesis on the physiology of the metabolism of a student of his colleague Leblond in Montreal in his Austrian journal *Biologia Generalis*.[35]

Bertalanffy's article in *Nature* also attracted the interest of the bio-mathematician Kostitzin, whose "mathematical biology", published in 1937, he knew well—it theorized in particular the phenomena of growth in the case of living organisms as well as in that of populations. Since Bertalanffy's writings were "completely unknown" in France where Kostitzin worked, he contacted him in order to obtain his writings and to include an analysis of them in the revised edition of his book[36] of 1937. Thus began a fruitful and regular correspondence between them, which lasted four years. Kostitzin was particularly interested in Bertalanffy's ideas; both shared the same admiration for Goethe and the same interest for mathematical theories of growth.[37]

9.3.4 The Beginnings of Bertalanffy's Research on the Pathology of Cancer

In connection with his theory of organic growth, Bertalanffy furthermore undertook research on cancer in Ottawa. It was financially supported in Canada from 1950

---

34 Letters from W. Ludwig to L. von Bertalanffy (10/10/1950) and from L. von Bertalanffy to W. Ludwig (22/11/1950), BCSSS archive.
35. Letter from L. von Bertalanffy to C.P. Leblond (26/10/1949), BCSSS archive.
36. Letter from V.A. Kostitzin to L. von Bertalanffy (21/03/1949), BCSSS archive.
37. Correspondence between L. von Bertalanffy and V.A. Kostitzin (21/03/1949—17/03/1953), in particular 25/12/1950, BCSSS archive.

onwards by the National Research Council and above all by the National Cancer Institute. Three articles on this problem were co-published by Bertalanffy between 1950 and 1953.

His research concerned first of all the study of the role of the ribonucleic acid (RNA) in the carcinogenesis and the influence of hormones in the formation of cancer. Then Bertalanffy began to develop a method for the diagnosis of cancerous pathologies on the basis of studies carried out in Vienna since the beginning of the century, especially by the botanist K. Höfler, whom he had got to know in 1943 (F. von Bertalanffy, 1973: 739-40). These studies were focussed on the technique of fluorescence microscopy, which combines the use of ultraviolet rays with colored fluorescent markers. Bertalanffy applied this technique to the study of cell growth. And he discovered that one marker, the acridine orange, colors the DNA of cells green, while coloring their RNA in shades ranging from auburn to an orange-like red, according to the quantity of RNA in the cell. As the amount of RNA in the cell depends on the synthesis of proteins and thus on the metabolism, the acridine orange thus allows to detect a metabolism and an abnormally elevated cytological growth rate. Bertalanffy believed that the use of this marker should therefore allow a rapid and economic diagnosis of cancer.

9.3.5 Development and Promotion of the General Systemology

In Ottawa Bertalanffy finally completed the "more detailed account" of his general systemology[38] (his article published in 1950 in the *British Journal for the Philosophy of Science*) and worked on its promotion. While regretting, nevertheless, that his obligations and his experi-

---

38. Letter from L. von Bertalanffy to F. Gessner (28/10/1949), BCSSS archive.

Chapter 9: Arrival in Canada: Montreal and Ottawa

ments did not leave him enough time for that purpose.[39]

From December 27 to 29, 1950, he participated in the 47th conference of the Eastern division of the American Philosophical Association, which took place in Toronto that year. Its topic was "Cybernetics and Teleology", and Bertalanffy was invited after Hans Jonas (1903-1993) had specifically recommended him to the organizer,[40] the philosopher Max Black (1909-1988). Bertalanffy and Jonas maintained a friendly relationship and met regularly at that time, probably at the home of the former. They exchanged in particular their views on mysticism, and Jonas needed Bertalanffy's opinion regarding a book he was writing on the history of gnosis.[41] Furthermore, and on Jonas's request, Bertalanffy did everything to get him a post in Ottawa[42] and succeeded in 1951.

At the conference in Toronto Bertalanffy found himself next to Jonas, next to William V.O. Quine (1908-2000), but also next to the neo-positivists Hempel and Ernst Nagel (1901-1985),[43] whom he had known in Vienna and met again on that occasion—his contact with Hempel had in fact been taken up again several weeks earlier in a correspondence. At this conference Bertalanffy outlined the ideas of his general systemology, and his two lectures on the subject gave rise to a constructive controversy that pitted him against Jonas and Hempel (who found this "theory" "highly stimulating", but denied it a systematic value and limited it to a heuristic

---

39. Letter from Bertalanffy to F. Keiter (09/11/1949), BCSSS archive.
40. Letter from H. Jonas to M. Black (23/02/1950), BCSSS archive.
41. Letter from L. von Bertalanffy to H. Jonas (01/04/1950), BCSSS archive.
42. Letter from L. von Bertalanffy to H. Jonas (09/03/1950), BCSSS archive.
43. Letter from L. von Bertalanffy to C.G. Hempel (06/11/1950), BCSSS archive. Brauckmann, 1997: 11-12.

function[44]). A third lecture held by Bertalanffy was not less important, since it defined the position of his general systemology in relation to cybernetics and introduced a dichotomy that determined not only his later work in psychiatry and psychology but afterwards ran more generally through the history of the systemic movement. Bertalanffy propounded that cybernetics is based on the concepts of feedback and homeostatic equilibrium. It develops the vision of essentially reactive systems, whose structural conditions are rigid and whose degree of organization can only increase when external "information" is provided: it is basically only a refinement of the behaviorist model of "stimulus and response", which it supplements with the concept of feedback. To this "mechanistic theory of systems" Bertalanffy opposed his general systemology as an "organismic theory of systems": its fundamental concepts are those of the "open system" (open to exchanges of matter and energy, and not only to "information") and of dynamic interaction; its emphasis is put on the states of non-equilibrium, more precisely on the steady state; and it develops the vision of primarily active systems, capable—due to their own dynamics—of organizing themselves progressively. Bertalanffy, perfectly in line with his "organismic" principles, conceived cybernetics as a particular field of general systemology: that which deals with mechanized systems whose characteristics of self-regulation rest only on fixed structural arrangements and which are thus "secondary" with regard to those "primary" ones, which result from the dynamic interactions between the components of the system.

During this (first) period in Canada, Bertalanffy's article, which contributed perhaps most to the diffusion of his general systemology, was published in November 1953 in the widely read journal *Scientific Monthly*. In this

---

44. Letter from C.G. Hempel to L. von Bertalanffy (02/12/1950), BCSSS archive.

paper, entitled "Philosophy of Science in Scientific Education", he relies on several recent calls for the creation of an "education of the scientific generalist" that emerged out of an awareness of the dangers of specialization for both science as such and for its social function. And he outlines the plan for a course of scientific philosophy in which his general systemology is supposed to meet such a need for integration. It is also a text in which Bertalanffy attacks logical positivism for the first time in a detailed manner, in particular the physicalist conception of the unity of science. And above all in which, again for the first time, he introduces the term "perspectivism" to describe his philosophy of knowledge (presented in opposition to positivism and empiricism) and draws its outlines without however systematizing it—while this philosophy, operating in fact in all his works since the late 1920s, had previously remained implicit. In this text he insists especially on the idea that science is a network of "freely chosen" conceptual constructions, "a conceptual order we bring into the facts", which is nevertheless not arbitrary; that it is a set of "perspectives" among others in the world, whose fruitfulness and efficiency depend on the capacity, based on the immanent logic of symbolic systems (what he called since 1951 their "algorithmic" character), to establish invariant relations in the phenomenal world to which the common sense could not have access; that this efficiency is proof that the invariants in question represent certain "traces" of an external reality to our consciousness; but that no scientific construction can for all that pretend to exhaust this reality.

9.3.6 The Aborted Project of the Philosophical History of Biology and the English Version of *Das Biologische Weltbild*

To the aforementioned group of works Bertalanffy completed in Ottawa, we may add one which he did not finish, probably because his priorities were elsewhere. When he was in London on November 10, 1948, he received a letter from the University Press in Bonn that asked him to send them his manuscript on the philosophical history of biology in the course of 1949,[45] which he had discussed with Rothacker in 1943. In a context in which such a contribution would have been financially welcome, Bertalanffy had contacted Rothacker again on the subject in 1947, without however pursuing it (Brauckmann, 1997: 8-9). Rothacker, who considered Bertalanffy's work "brilliant" but never saw a manuscript arriving, reminded him of the subject on July 25, 1949, while advising him to link up with Ungerer, who had already demonstrated his competence to carry out such a historical work.[46] The text in question, titled *Geschichte des Lebensproblems* ("History of the Problem of Life"), was supposed to constitute the first volume of an editorial project entitled "Histories of Sciences in Documents". Bertalanffy discussed this with Gassner and drafted an outline for it. It was not a history of biology but a history of the epistemological and ontological problems of biology that aimed at showing how Aristotle's question of the difference between living and non-living beings was posed in the present.[47] But even though he accepted the project, Bertalanffy very quickly showed hardly any concern about meeting deadlines and was satisfied above all with working on this history for his personal pleasure[48] so that the project, unfinished until 1953, was not completed with him. It was finally Ungerer who (brilliantly) took charge of this task.[49]

---

45. Letter from *Universitäts-Verlag Bonn* to L. von Bertalanffy (10/11/1948), BCSSS archive.
46. Letter from E. Rothacker to L. von Bertalanffy (25/07/1949), BCSSS archive.
47. Letter from F. Gessner to L. von Bertalanffy (08/09/1949); letters from L. von Bertalanffy to T. Ballauf (13/09/1950) and (03/11/1950), BCSSS archive.
48. Letter from L. von Bertalanffy to F. Gessner (28/10/1949), BCSSS archive.
49. E. Ungerer, *Die Wissenschaft vom Leben—Eine Geschichte*

Still in Ottawa, by contrast, Bertalanffy finalized the English edition of *Das biologische Weltbild*, though not without difficulties with the translator. It appeared under the title *Problems of Life* in 1952 and received a number of very favourable reviews in America and contributed a bit more to making Bertalanffy known there. He also supervised from a distance the reissue of the second volume of his *Theoretische Biologie* in Switzerland, which appeared in 1951. But in May 1950 he had legal concerns in this respect: he was forced to start proceedings against an American publishing house (J.W. Edwards, Ann Arbor), which had just published and put into circulation a reprint of this book under the licence of the U.S. government without the authorization of the Swiss publishing house that possessed the corresponding rights.[50]

9.3.7 A Series of Conferences in the United States: Meeting Aldous Huxley

Following several invitations, Bertalanffy undertook a series of fifteen lectures in the United States in September and October 1952.[51] In the midst of the McCarthy era he was first denied a visa because of his involvement in the weekly newspaper *Die Woche* in post-war Vienna (which was circulated by the Globus-Verlag, a publisher of the Austrian Communist Party, KPÖ). He only received his visa after his wife had blackmailed the American embassy in Canada, threatening to stir up the press

---

der Biologie, Orbis Academicus, Freiburs/München, K. Alber, 1966. This book can be found amongst the remains of Bertalanffy's library.
50. Letter from L. von Bertalanffy to V.A. Kostitzin (20/04/1950), BCSSS archive.
51. Letter from L. von Bertalanffy to J. Alexander (19/06/1952), BCSSS archive. Report by L. von Bertalanffy to the Canadian Cancer Institute, 1950-1953, Archive of the University of Ottawa.

and the American universities where he was supposed to speak (Davidson, 1983: 61).

His lectures took place at the University of Chicago (on the invitation of the psychiatrist Roy R. Grinker (1909-1975)); at the University of Texas (at the medical division of Galveston on the invitation of the biochemist Wiktor W. Nowinski, at the main division in Austin on the invitation of Weiss and the biochemist Chauney D. Leake, and in Houston); in Topeka, Kansas (invited by Menninger); at the Washington University in Saint-Louis, Louisiana; at the medical school of Denver, Colorado; and in Stanford and Berkeley, California (on the invitation of the psychologist Egon Brunswik (1903-1955)—a former student of Karl Bühler—but also of Reichenbach, whom he met again on that occasion).[52]

During his journey in California Bertalanffy met Aldous Huxley (1894-1963) for the first time in person, with whom he had maintained correspondence since his arrival on the American continent in 1949. Their relationship, which would last until Huxley's death, was friendly and marked by great mutual admiration.[53] Bertalanffy admired in Huxley his "Goethean universality"[54] and considered him "the only universal thinker" since the death of Gerhart Hauptmann[55] (1862-1946), winner of the Nobel Prize in literature in 1912. As to Huxley, he thought Bertalanffy exceptionally open, intelligent and cultivated.[56]

---

52. L. von Bertalanffy: Activity report to the National Institute of Cancer (1950-1953), Archive of the University of Ottawa.
53. Correspondence between L. von Bertalanffy and A. Huxley, in Gray and Rizzo, 1973: 187-211. Davidson, 1983: 61-62.
54. Letter from L. von Bertalanffy to A. Huxley, in Gray and Rizzo, 1973: 188.
55. Letter from L. von Bertalanffy to F. Gessner (28/10/1949), BCSSS archive.
56. Letter from A. Huxley to A. Watts (17/10/1952), in Gray

9.4 Bertalanffy's Malaise in Ottawa and, More Generally, in America: An Aristocrat among Philistines

This journey to the United States was for Bertalanffy a breath of fresh air, as he did not like Ottawa and complained about it from 1951 onwards.

He admitted that his research was supported more than was necessary by the Canadian institutions, considered his working conditions "pretty good", the scientific productions of his department satisfying and his life "comfortable".[57] But he had no assistant who suited him and always tried hard to recruit one.[58] In addition, the libraries in what he considered a "small university" were a big problem for him: he could only obtain European books and articles with difficulty.[59] Above all, the University of Ottawa was "run by a Catholic order and at the mercy of the whims of ignorant and arrogant priests" who confronted him with all sorts of obstacles in his will to introduce teaching that broke with an educational system that, according to him, made it impossible to find meaning in what was acquired and that was in fact limited to the aim of delivering a diploma.[60] He did not

---

and Rizzo, 1973: 195.
57. Letter from L. von Bertalanffy to A. Mittasch (20/04/1949); letters from L. von Bertalanffy to F. Gessner (28/10/1949 and 12/11/1951); letter from L. von Bertalanffy to J. Alexander (02/04/1952); letter from L. von Bertalanffy to A. von Auersperg (15/05/1952); letter from L. von Bertalanffy to E. Brunswik (28/04/1952); letter from L. von Bertalanffy to C. Singer (30/041952), BCSSS archive.
58. Letter from L. von Bertalanffy to B. Grad (07/01/1951), BCSSS archive.
59. Letter from L. von Bertalanffy to H. Friedmann (12/01/1951), BCSSS archive.
60. Letter from L. von Bertalanffy to J. Alexander (02/04/1952); letter from L. von Bertalanffy to A. von Auersperg (15/05/1952); letter from L. von Bertalanffy to

put up with the "intrigues" within the university either, provoked by the policy of its management. He ultimately believed that Ottawa consumed "much of [his] time and energy" and that he "deserve[d] a better place".[61]

Like his old Viennese friend Charlotte Bühler, who also emigrated to America, Bertalanffy experienced in fact serious difficulties integrating into an academic world committed to egalitarian and anti-authoritarian values.[62] He despised his fellow biologists, in his eyes just about able to "talk about the weather" when they do not fuss over their small specialized research field, and incapable of talking about philosophy or literature.[63] This contempt for biologists was more general, since he thought that the scholars in this discipline were "especially slow on the uptake" as soon as the questions became theoretical, which, according to him, explained the time his ideas needed to become established in this field.[64] Bertalanffy was in fact fundamentally convinced that the systemic point of view was only accessible to "a small elite"[65] and that his destiny was to be "always one step ahead of other

---

Brunswik E. (28/04/1952); letter from L. von Bertalanffy to F. Gessner (28/10/1949); letter from L. von Bertalanffy to C. Singer (30/04/1952), BCSSS archive.
61. Letter from L. von Bertalanffy to C. Singer (30/04/1952); letter from L. von Bertalanffy to J. Alexander (02/04/1952), BCSSS archive.
62. Letter from P. Lazarfeld to R. Tyler (16/05/1956), attached to a letter from C. Bühler to L. von Bertalanffy (20/05/1956), BCSSS archive.
63. Letter from L. von Bertalanffy to M. von Auersperg (29/01/1950), BCSSS archive. Translated from the German by ES.
64. Letter from L. von Bertalanffy to F. Gessner (28/10/1949), BCSSS archive. Translated from the German by ES.
65. Letter from L. von Bertalanffy to F. Keiter (09/11/1949), BCSSS archive. Translated from the German by ES.

Chapter 9: Arrival in Canada: Montreal and Ottawa

people", which he believed to be the reason why he fit in so poorly among orthodox scientists.[66]

Bertalanffy's difficulties with integration were also attributable to his personality. Colleagues who knew him well and understood his thinking certainly described him as "timid" and "nice", radiating a "simple kindness", an "openness of the mind" and unusual "dignity". But they also agreed on judging him an "eccentric" and a "complicated" person who suffers from "mannerism" and very often leaves a bad impression on most of his colleagues, especially when they are not willing to listen to his ideas (Hammond, 2003: 107).

Beyond the academic world, Bertalanffy experienced difficulties in adjusting to an America whose culture he had despised since his arrival. He despised not only the stupidity of popular magazines and the omnipresent mercantilism, but more deeply the "American mentality", which embodied for him a "period of levelling and vulgarity". He missed the "atmosphere of the old culture",[67] and living in America he felt like living in a "cultural and spiritual desert".[68] Subject to recurrent phases of "deep depression",[69] Bertalanffy lived in fact with the profound and aristocratic conviction to have been "born first into a lousy, then a great, then an iron

---

66. Letter from L. von Bertalanffy to F. Gessner (12/11/1951), BCSSS archive. Translated from the German by ES.
67. Letter from L. von Bertalanffy to M. von Auersperg (29/01/1950), BCSSS archive. Translated from the German by ES.
68. Letter from L. von Bertalanffy to A. Mittasch (14/06/1952); letter from L. von Bertalanffy to A. von Auersperg (15/05/1952), BCSSS archive. Letter from L. von Bertalanffy to E. Rothaker (15/07/1953), in Brauckmann, 1997: 12-13. Translated from the German by ES.
69. Letter from L. von Bertalanffy to A.H. Bentley (20/09/1950), BCSSS archive. Translated from the German by ES.

and finally a miserable era".[70] It is doubtlessly a sign of his growing malaise that in this period, more precisely in 1953, his supreme contempt for mass culture and the "American mentality" went beyond his private correspondence and settled in at the heart of his publications, where he castigated an utilitarianism that degrades the value of science and enslaves it to political and economic reasoning, leading to an in his eyes intolerable situation, where a Nobel Prize has less value than an underage boxing champion, a television star or a fashion model (L. von Bertalanffy, 1953: 238).

Bertalanffy thought so much about emigrating that he proposed to his old friend, the former SS physician von Auersperg, on May 15, 1952, to go with him to Chile and to develop an institute of general biology or of biological medicine there, on the condition that financing was not a problem[71]—but this condition was not fulfilled. He also wrote to Rothacker on July 15, 1953, to complain about the American cultural destitution and to confide to him that "if he was offered a nice chair in Germany, he would certainly take it into consideration"[72]—his letter remained unanswered.

9.5 Bertalanffy's Conflicts with the University of Ottawa (1)

Even though Bertalanffy's unrest largely surpassed the framework of the University of Ottawa, his desire to leave at least the university was above all due to his conflicts with its administration. It was rooted in a dispute that emanated from a discussion with Danis on

---

70. Letter from L. von Bertalanffy to F. Gessner (19/09/1951), BCSSS archive.
71. Letter from L. von Bertalanffy to A. von Auersperg (15/05/1950), BCSSS archive.
72. Letter from L. von Bertalanffy to E. Rothacker (15/07/1953), in Brauckmann, 1997: 12-13.

Chapter 9: Arrival in Canada: Montreal and Ottawa

May 6, 1950. This conflict was connected to the informal nature of Danis's initial promise that Bertalanffy would obtain the direction of the Department of Biology when its director (Le Bel) retired. On that day Danis actually let Bertalanffy understand that the question was not settled, which caused Bertalanffy to write a first indignant letter in which he demanded a formal assurance of obtaining the post.[73] As there was no reply, the same request was sent again one month later,[74] but it did not have any effect either. On December 5 Bertalanffy suggested to the dean of the university (A.L. Richard) that the Department of Biology should be split and that a "Department of Biological Research and Experimental Medicine", with himself as head of the department, should be created.[75] But the proposal remained again unanswered.

The relationship between Bertalanffy and the administration of his department and of the university in general deteriorated significantly as of April 1952, which explains that his strong wish to have a change of scene manifested itself from this period onwards. On April 7 Bertalanffy learned that arrangements for the appointment of a member of the religious congregation that directed the university as head of the Department of Biology after Le Bel's departure had been made and that, above all, the plan was to separate teaching and research in the new buildings of the Faculty of Medicine that was then under construction. Bertalanffy made a complaint to Dean Richard about this on April 21, but nevertheless accepted the intended appointment under two conditions: that his request of December 5, 1950, should be acceded to and that his salary should be raised. He demanded furthermore an extension to the building of

---

73. Letter from L. von Bertalanffy to A.L.M. Danis (06/05/1950), Archive of the University of Ottawa.
74. Letter from L. von Bertalanffy to A.L.M. Danis (02/06/1950), Archive of the University of Ottawa.
75. Letter from L. von Bertalanffy to A.L. Richard (05/12/1950), Archive of the University of Ottawa.

biological research in relation to what had been provided for in the new buildings.[76] This last request was not only, apart from the odd detail, met with a point-blank refusal, but also with a call to order concerning the employment of his wife, which should have been temporary and which had in fact become permanent.[77] Still not receiving any assurance regarding the conditions that he had set on April 21, Bertalanffy wrote to the Rector of the University of Ottawa (the Reverend Rodrigue Normandin) on December 6 in order to set four exclusive conditions of which one had to be satisfied if they wanted him to stay at the university: that he succeeded Le Bel as head of the Department of Biology, that his proposal of December 5, 1950, was accepted, that he became head of the Department of Physiology (a post that was vacant at that time), or else that he became head of the Department of Biophysics, whose creation had already been provided for.[78] The rector attempted to appease him by answering on January 8, 1953 that even if Father Danis had no authority to promise Bertalanffy Le Bel's position, nobody else than him was nevertheless considered for the post.[79]

But his relations kept deteriorating. He was told that he did not put enough energy into his teaching and devoted himself only to his research. He complained on his part to the rector on October 22, 1953, saying that his salary was very much inadequate considering his status and his international reputation, denouncing the fact that his colleagues of equal rank in the same university earned up to 50% more than he did. He demanded not only a substantial increase of 20% but also its retroactiv-

---

76. Letter from L. von Bertalanffy to A.L. Richard (21/04/1952), Archive of the University of Ottawa.
77. Rectification of Dr Bertalanffy's letter (1952, not dated), Archive of the University of Ottawa.
78. Letter from L. von Bertalanffy to R. Normandin (06/12/1952), Archive of the University of Ottawa.
79. Letter from R. Normandin to L. von Bertalanffy (08/01/1953), Archive of the University of Ottawa.

ity as of July.[80] And two weeks later he repeated the same request in an outraged tone and attached a document which showed that his current salary was lower than that of a nursery school teacher.[81] The rector answered him by saying that this letter was insulting to him and by giving him the order to invest himself more in his teaching. As regards the increase of his salary, he told Bertalanffy that he needed to pass through the correct channels for such a request and at the same time refused this increase by arguing that the budget for the academic year was not to be touched and that the university was in financial difficulties.[82]

But while his destructive relations made Bertalanffy's future in Ottawa uncertain, many other prospects, which would satisfy his growing desire to leave Ottawa, presented themselves simultaneously. His journey to Chicago in 1937-1938 would help him to have yet another change of scene.

9.6 An Invitation to Stanford (California): The Prospect of an Institutionalization of the General Systemology

9.6.1 Bertalanffy's Affinities with the Committee on the Behavioral Sciences of Chicago

Even if he had to suffer an adverse decision from the neo-positivist side, his project of general systemology had in fact introduced him positively at the University of Chicago during that time: he had

---

80. Letter from L. von Bertalanffy to R. Normandin (22/10/1953), Archive of the University of Ottawa.
81. Letter from L. von Bertalanffy to R. Normandin (09/11/1953), Archive of the University of Ottawa.
82. Letter from R. Normandin to L. von Bertalanffy (22/10/1953), Archive of the University of Ottawa.

contributed to the idea of a *Committee on the Behavioral Sciences*, founded in this university in 1949 by James G. Miller (1916-2002). Miller's work, just like that of Gerard, who joined him in this project, was then driven by the unceasing research for a unique pattern that permitted to integrate biology and the social sciences and to develop applicable methods for both. His department and the expression "behavioral science" that he also coined in 1949, express a dissatisfaction with the fragmentation of the disciplines that study man and human societies and embody a will to integrate biological, psychological and social dimensions of human behaviour in a single study. Miller and Gerard joined forces in this regard with the mathematician Anatol Rapoport (1911-2007), who also worked in Chicago at that time. The latter was especially interested, just like Rashevsky, with whom he worked, in the application of mathematical models to biological and social phenomena. He was convinced that analyses of the form and the content of a phenomenon could be independent and that mathematical models of formal structures were consequently bound to provide a powerful means of theoretical unification.

One of Bertalanffy's lectures in the United States in 1952 had followed an invitation to Chicago by Roy Grinker, one of Miller's colleagues committed to promoting the same ideas. This invitation had allowed Bertalanffy to contact the members of the *Committee*, who considered his general systemology promising. He also met Gerard there again, who provided him moreover with a letter of support on December 5, 1953, when he applied for a grant from the Guggenheim Foundation (Hammond, 2003: 144-145).

9.6.2 Making Contact with Kenneth E. Boulding and Joining Forces in View of an Institutionalization of the general systemology

Also in November 1953 Bertalanffy started a correspondence with the heterodox economist Kenneth E. Boulding (1910-1993), who thought that there was no autonomous science that one could call "economy", because it was—like all social sciences—only a point of view on a single subject, the social system, for the understanding of which an interdisciplinary work of integration had to be developed. In 1953 he outlined a general theory of organizations that discusses the similarities between living organisms and social organizations and shows how the behavior, the growth and the survival of organizations in general is determined by their internal structure, sketching out in particular a theory of the structural limits of growth.

Boulding, who gave a seminar at the University of Michigan on these problems, noted the parallelism of his ideas with those of Bertalanffy when discovering his article published in November 1953 in *Scientific Monthly* (Hammond, 2003: 217, 246)—the same article had also attracted the attention of Grinker, who, after having read it, invited Bertalanffy to work in Chicago in a research group he directed, entitled "Unified Theory of Human Behavior".[83] Thus Boulding took the initiative and started a correspondence with Bertalanffy, in which they discussed the possibility of creating an institute dedicated to the development of a general systemology, an idea about which Bertalanffy had already talked with Huxley in 1952, who had encouraged him to procure financial support from the Ford Foundation for this purpose. In early 1954 Bertalanffy and Boulding sent a letter to various scholars whom they believed to be interested in such

---

83. Letter from L. von Bertalanffy to A. Huxley (13/01/1954), in Gray and Rizzo, 1973: 202.

a project, notably in Chicago. They received enough enthusiastic answers to be encouraged in this direction, in particular by that of Rapoport, who wrote to Bertalanffy to that effect on June 7 (Hammond, 2003: 217, 246).

9.6.3 Bertalanffy within the Context of the Foundation of the Center for Advanced Study in the Behavioral Sciences (CASBS)

The *Committee on the Behavioral Sciences* directed by Miller formed the spearhead of the realization of a project that the Ford Foundation accepted to finance, because it agreed with its commitment to promote peace, democracy, education and economic welfare on the basis of a social science capable of providing the necessary knowledge for these purposes: the creation of a "Center for Advanced Study in the Behavioral Sciences (CASBS)". This centre was intended to unite experts coming from different academic fields and thus to promote the "interdisciplinary integration and exchange in behavioral sciences" in view of establishing the determining factors of the latter.[84]

In the spring of 1954, when the opening of the CASBS took place at Stanford University, California, Boulding contacted his designated director Ralph W. Tyler (1902-1994), equally the director of the division of social sciences in Chicago, to recommend Bertalanffy as one of the few researchers (thirty eight) who had the privilege to launch the first year of operation—Boulding was himself already among them (Hammond, 2003: 217).

At the same time, precisely on March 20, 1954, Bertalanffy contacted Rothacker in Germany again in order to inform him that he would be very interested in his succession, as Rothacker was becoming Emeritus Pro-

---

84. Hammond, 2003. Preliminary report from L. von Bertalanffy to the *Center for Advanced Study in the Behavioral Sciences* (17/12/1954), BCSSS archive.

Chapter 9: Arrival in Canada: Montreal and Ottawa

fessor on April 1. He hoped in fact that the chair of philosophy of the latter could be transformed into a chair of natural philosophy [*Naturphilosophie*], having therefore the competence to succeed him (Brauckmann, 1997: 13). Bertalanffy thus planned a return to Europe in any case: whether after a period of working at the CASBS or even more in case he was not invited to work there. His letter remained unanswered, but he repeated a request in that direction on May 24, again without response, stating that he was "more engaged in philosophical questions than is commonly known", a fact which he had "for a long time carefully hidden in order not to provoke Boeotian biologists, who liked to call him a philosopher".[85]

Bertalanffy left for the United States in July, not only to co-direct a conference in Milwaukee and to speak there, but also and above all to meet the founders of the CASBS in Chicago and to discuss the organization of its first year of operation.[86] Previously at the University of Ottawa he had applied for a leave of absence with two thirds of his salary for the period of September 1, 1954 to January 31, 1955. The reason he gave for this request was that he wanted to finish several books and articles that were in preparation. His request was accepted and all necessary arrangements were made.[87]

---

85. Letters from L. von Bertalanffy to E. Rothacker (20/03/1954 and 24/05/1954), in Brauckmann, 1997: 13. Translated from the German by ES.
86. Letter from L. von Bertalanffy to P. Gendron (26/07/1954), Archive of the University of Ottawa.
87. Letter from L. von Bertalanffy to S.F.M. Wotherspoon (28/09/1954), BCSSS archive. Letter from S.F.M. Wotherspoon to Vincent and Addy (18/10/1954), Archive of the University of Ottawa.

9.7 Bertalanffy's Conflicts with the University of Ottawa (2)

Even though Bertalanffy saw promising prospects opening in the United States and was invited to the CASBS, the horizon darkened more and more for him in Ottawa.

9.7.1 The Beginning of an Open Crisis: Bertalanffy Overtaken in the Run for Head of the Department of Biology

On July 15, 1954, the rector announced to him that Jean R. Baudry, from the University of Montreal, was appointed to succeed Le Bel as head of the Department of Biology. He was furthermore notified that his wife would no longer be employed by the university as of September 1.[88]

Bertalanffy, who discovered the letter in question on his return from Chicago, sent a letter to the dean (Pierre Gendron) on July 26, in which he voiced his "extreme surprise", but told him above all that such an appointment represented an ingratitude towards him, given his "loyal services" and the international stature the department of research had acquired within the four years in which he had organized and directed it, a stature which was essentially owing to his efforts. He attached to his letter a long list of quotations praising the quality of his work.[89] Naturally, this letter did not change the situation. But Bertalanffy's trials with the University of Ottawa were far from being over.

---

88. Letter from R. Normandin to L. von Bertalanffy (15/07/1954), Archive of the University of Ottawa.
89. Letter from L. von Bertalanffy to P. Gendron (26/07/1954), Archive of the University of Ottawa.

Chapter 9: Arrival in Canada: Montreal and Ottawa

9.7.2 The Opposition of the University of Ottawa to Bertalanffy's Departure for Stanford

In July 1954 Bertalanffy received his official invitation to the CASBS for the period from October 1, 1954, to September 1, 1955. In October he thus made a request for an adequate extension of his leave of absence which, initially granted for five months, would this way be extended to twelve months. He gave as a reason the honour which this invitation represented for the University of Ottawa, as only distinguished scholars were invited to the CASBS and since he was the only representative from Canada who had been invited—his request for obtaining the Canadian Nationality had also been acceded to several weeks earlier. He offered in exchange not to be paid by the University of Ottawa throughout this period, in contrast to the initial agreement concerning the leave of five months. And he made all the necessary proposals so that his additional absence did not disturb the organization of the department.[90]

But Baudry, who had been appointed as head of the Department of Biology in his place, made a complaint to the dean of the Faculty of Sciences and listed the difficulties that this leave of absence would raise: the supervision of doctoral candidates, the management of funds allocated to Bertalanffy by external institutions and the probability that they would not be renewed, as well as the replacement for his teaching. And he recommended that Bertalanffy's request be refused.[91] The dean followed him and informed the latter of this refusal on September 7. As a reason he gave in particular a too short period for

---

90. Letter from L. von Bertalanffy to S.F.M. Wotherspoon (28/09/1954), BCSSS archive. Letter from J.R. Baudry to P. Gendron (02/09/1954), Archive of the University of Ottawa.
91. Letter from J.R. Baudry to P. Gendron (02/09/1954), Archive of the University of Ottawa.

the university to take necessary steps[92]—thereby ignoring almost five months of already granted leave, during which these measures could have been taken.

Bertalanffy reacted on September 14 by sending a letter to the dean demanding that his case should be reconsidered. Legally on leave since September 1, he left Canada on the 14th to go to California with his wife, not without having previously contacted a lawyer. Officially, i.e., according to him through his lawyer, he intended to return at the end of his leave, on January 31, in case the decision of the university would be held up.

9.7.3 A Welcome Unlawful Dismissal and Its Advantageous Transformation

Bertalanffy had arrived in Stanford when, on September 23, through the lawyers of the University of Ottawa, he received a letter which filled him with consternation without making him in the least sad: the university notified him of his dismissal, "in view of his decision to leave his employment for eleven months, after leave of absence for that period has been refused". The university thought in fact that his request for re-examination together with his departure for California amounted to a refusal to accept the decision it had taken and thus forgot about the leave of five months it had already granted him.

Bertalanffy reacted as soon as he received this letter by informing his lawyer of it on September 28. He did "not care to go back to Ottawa" but did not intend to "make it so easy for them" by admitting it. On the contrary, he drew on the complete illegality of this dismissal, which was without foundation (he had never notified, neither in writing nor by word of mouth, his intention to be absent beyond the month of January) and belied the very regulation of the university (three months' notice

---

92. Letter from P. Gendron to L. von Bertalanffy (07/09/1954), Archive of the University of Ottawa.

before the beginning of the term were required), to tell the university that he would drop the charges and would agree to resign if he were paid eight months of his full salary.[93] But an arrangement was finally found without having to take the case to court. Backed by his lawyer, Bertalanffy dropped all his charges on October 18 under the condition that the university accepted his resignation as from September 30, 1954; that it undertook to specify in writing that his departure was on his own initiative and did not result from any misdemeanor; that it abandoned all claims against him; and finally, that he received his complete salary for the month of September.[94] The university accepted his conditions. The official letter of the rector notifying Bertalanffy that his resignation as from September 30 had been accepted was sent to him on November 17, 1954.[95]

---

93. Letter from L. von Bertalanffy to S.F.M. Wotherspoon (28/09/1954), BCSSS archive.
94. Letter from S.F.M. Wotherspoon to Vincent and Addy (18/10/1954), Archive of the University of Ottawa.
95. Letter from R. Normandin to L. von Bertalanffy (17/11/1954), Archive of the University of Ottawa.

CHAPTER 10
The First Emigration to the United States and the Society for General Systems Research (1954-1955) at an Embryonic Stage

10.1 Bertalanffy's Theoretical Activities in Stanford

The eleven months which Bertalanffy spent in Stanford were full of fruitful discussions for him, during which his project of general systemology was updated and reached the early stages of institutionalization, in which a gradual reorientation of his work towards psychology and psychiatry took place and in which we find a return to philosophy of culture. It was also a period in which his perspectivist philosophy of knowledge became fully explicit and reached its maturity.

10.1.1 The Seminars of the CASBS in General and that of "General Systems" in Particular

The thirty-eight visiting scholars during the first year of operation of the CASBS were only there for this period, as a principle of annual rotation was constituent element of the centre. But very rapidly a dynamics of totally interdisciplinary exchanges developed, which, for the researchers in question, opened up many paths for their future work. They split up into seminars (nineteen at the beginning) and each of them could participate in several of them.

Gerard and Rapoport, in partnership with Paul Lazarsfeld (1901-1976) and Duncan Luce, worked for instance on the problem of theorizing social networks by drawing formal analogies with the network of neurons. Rapoport worked also with Alex Bavelas (a student of Lewin who had mathematized group dynamics) and Stephen Richardson (who contributed to the unpublished work of his father on the mathematical analysis of the arms race). Gerard and Franz Alexander (1891-1964)—who had worked with Freud—collaborated in order to develop a synthesis of concepts coming from neurophysiology and psychoanalysis (Hammond, 2003: 246). As regards Bertalanffy, he worked with the anthropologist Raoul Naroll (1920-1985) on the interpretation of certain social phenomena on the basis of the mathematical principle of allometric growth, which he had already widely used in the study on relative organic growth.[1] It was an important work for Bertalanffy, which constituted probably the clearest example of demonstrating the fertility of his general systemology and of how he understood its application.

This work of Bertalanffy and Naroll was in fact in line with a weekly seminar taught by Bertalanffy and Boulding and entitled "General Systems". In this seminar Bertalanffy and Boulding, followed by Gerard and Rapoport, gave the introductory lectures in order to define its spirit. In particular Bertalanffy's lecture, which outlined an overview of his conception of general systemology, was published in March 1955 and would constitute his most cited article on the subject. Three future systems theoreticians of repute participated also in the seminar: Herbert A. Simon (1916-2001), Charles MacClelland and Lancelot Whyte. The topics they discussed were very diverse: in addition to the principle of allom-

---

[1]. Preliminary report to the CASBS by L. von Bertalanffy and R. Naroll on the seminar "General systems" (17/12/1954), BCSSS archive.

etry in biology and the social sciences they discussed the foundations of physics, the concept of homeostasis in biology, psychology and economy, the Malthus problem (exponential growth and its limits), the theory of administration and international relations, or the famous and very controversial "hypothesis" of the ethnologist and linguist Benjamin Lee Whorf (1897-1941) concerning the linguistic relativity of categories (Hammond, 2003: 247).

10.1.2 The Systematized "Perspectivist" Philosophy

Bertalanffy got to know this last "hypothesis" through Huxley, who had recommended him to read it.[2] The debates that sparked off from it prompted him, in the extension to his 1953 article in the *Scientific Monthly*, to summarize the reflections he had formulated in a scattered manner in his previous writings and to systematize the philosophy of knowledge which underlies his work on organic growth as well as his theory of open systems and his general systemology, all marked both by an advancement of the concept of "model" and by a rejection of the relativism it implied with other authors. In "An Essay on the Relativity of Categories", published in 1955, he expounded the foundations of his "perspectivism" which, while prefiguring "post-modernism", laid down the basis of its critique and of its overcoming even before its advent.

This "perspectivism" was in fact essentially created in the late 1920s and the early 1930s, after having read the works of Spengler and Jakob von Uexküll (1864-1944), and more even of various neo-Kantian philosophers. Bertalanffy had indeed provided the essential ideas in 1937 in a passage in *Das Gefüge des Lebens*, which his

---

2. Letter from A. Huxley to L. von Bertalanffy (10/10/1952), in Gray and Rizzo, 1973: 193.

article of 1955 largely developed further. His "perspectivism" is opposed to empiricism as well as to rationalist dogmatism (whether Cartesian or Kantian), as it affirms a biological and cultural determination of categories. He opposed to the postulate of a separation of the subject and the object of knowledge that of a construction of the latter which is rooted in the particular psychophysical organization of man and the modalities of which relate to culture. Thus man can only have "perspectives" in the world, and in this sense all scientific constructions are "models" that have limited validity. But simultaneously, this "perspectivism" rejects a radical relativism that would affirm equal value of all "perspectives". The symbolic capacity, which is peculiar to man and of which mathematics is the most refined expression, offers him, thanks to the "algorithmic" character of the systems it creates, the power to gradually "de-anthropomorphize" his concepts and models. It highlights the relational invariants at the heart of the phenomena, of the convergences of "perspectives" that are embodied in universal constants of the physical future. Bertalanffy, in full agreement with Cassirer, thus developed a modern version of the perspectivisms of Cusa and Leibniz: man, as he is a part of it, can only think "wholeness" in oppositions and can thus not grasp the absolute truth. But "wholeness" diffracts itself in each of its parts and man, in particular, can partially apprehend truths that transcend his biological and cultural determinants through his models.

With this perspectivism, Bertalanffy outlined, similarly to Konrad Lorenz (1903-1989), who was also inspired by his earlier writings, an "evolutionist theory of knowledge" that bears the seal of the pragmatism of Vaihinger and Reininger: the categories are the fruit of an interaction between man and his environment, which imply not only an adaptation to the latter but also his transformation. That they allowed him to survive and, to a certain extent, to control this environment, is the very proof that, despite their biological and cultural determi-

10.1.3 The Critique of the Social Role of the "Behavioral Sciences" and a New Moment for the "Dialectical Tragedy": The Individuality of Man as an "Ultimate Precept"

When completing this essay, Bertalanffy worked also at the CASBS—a particularly favorable context in this respect—on a development and a translation into English of a part of his contribution to the Alpbach symposium in 1947: that which was meant to be an outline of a philosophical anthropology, defining the specificity of man by his quality as a "symbolic animal". The article that followed, published in January 1956 and entitled "A Biologist Looks at Human Nature", marks on the one hand his desire to take up the project that should form the second volume of *Das biologische Weltbild* again; and on the other hand that of orienting himself more resolutely towards that part of the American academic world which revealed itself as the most receptive to his systemic thought: the "behavioral sciences" and, more generally, the so-called "human" sciences. This article aroused also great interest on the part of some anthropologists.[3]

Compared to 1947, the only true novelty introduced in this article in view of anthropology as such consisted in a definition and a clear characterization of symbolism by three criteria: "symbols can be defined as signs that are freely created, represent some content, and are transmitted by tradition". Bertalanffy tried to show that these criteria were sufficient to distinguish symbolism, the human language in particular, from all other forms of animal behavior, thus ensuring the uniqueness of humans.

---

3. Letter from R. Carneiro to L. von Bertalanffy (17/02/1956), BCSSS archive.

But besides this development, two new topics emerged in this article. They can essentially be interpreted as a response to a context in which all thoughts of "holistic" inspiration were attacked (by Karl Popper (1902-1994) and Hayek for instance) because of their alleged inherent tendency to generate and justify a political and social philosophy of totalitarianism. It was also a "re-orientation" characteristic of the history of "holistic" thinking in the twentieth century, which Bertalanffy shared in fact with most of the German-speaking philosophers of "wholeness", compromised in the legitimization of National-Socialism and expressing themselves after the war—such as Meyer-Abich.

Bertalanffy insisted first on the fact that to admit the biological foundations of human behavior, thus the rootedness of "behavioral sciences" in biology, was a necessity that needed to be supported with a clear challenge of biologism. He thought that this form of reductionism was even more dangerous than physicalism, as the individual cannot matter in such a perspective but only the "supraindividual whole"; so that it "logically leads to the theory and practice of a master race, total war and the totalitarian state". This warning conveyed for the first time in Bertalanffy's publications his keen awareness of the "tragedy of the concept of wholeness", whose existence his correspondence had revealed for several years already.

It is coupled with an unprecedented critique of the social role of the "behavioral sciences", which appears also at the end of his famous article on general systemology published several months earlier. Not only did this critique open a thorough attack on behaviorist psychology, which would from then on become a constant, but it also constituted partly an implicit critique of certain tendencies found among some members of the CASBS or other scholars who supported the centre with their activities, such as Miller and especially Gerard. Sharing the spirit that had prevailed at the foundation of the CASBS, Bertalanffy thought that the development of "behavioral sciences" was made necessary by the complexity of modern society and by the discrepancy between knowledge and the control of physical and biological nature on the one hand and between knowledge and the control of purely human problems on the other. But in the wake of Huxley he attacked the tendency, which he regarded as a modern version of the *Republic* by Plato, to make of the "behavioral sciences" the basis of "scientific totalitarianism", which, resting on an even more dangerous "psychological and social technology" (L. von Bertalanffy, 1955: 82) than the physical technology, acquires an unprecedented power of domination over the individual. These sciences had for him a considerable responsibility in the modern world: they can either work, as behaviorism does in his eyes, on building a mass society that denies the specific quality of the "symbolic animal" from which humans draw their dignity in order to substitute it with "a return to the conditioned reflex" through modern methods of propaganda; or, on the contrary, teach both what human society and human behavior have in common with other organizations and in how far they are unique.

Thus the pendulum motion of the "dialectical tragedy of the concept of wholeness" came back in Bertalanffy's publications at an authentically critical moment and had formed, beyond its ideological deviations between 1934 and 1935, the heart of his systemic philosophy since his doctoral thesis: one in which his philosophy is constructed around a perspectivist relation between "individual" wholeness and "supra-individual" wholeness, which rejects biologism and rests on the belief in the "metaphysical dignity of the individual".[4] After having mixed his "organismic" biology with the legitimization of the totalitarian state, Bertalanffy writes in a similar vein:

---

4. L. von Bertalanffy, 1926b: 45. Translated from the German by ES.

Man is not only a political animal, he is, before and above all, an individual. The real values of humanity are not those which it shares with biological entities, the function of an organism or a community of animals, but those which stem from the individual mind. Human society is not a community of ants or termites, governed by inherited instinct and controlled by the laws of the superordinate whole; it is based upon the achievements of the individual, and is doomed if the individual is made a mere cog in the social machine. This, I believe, is the ultimate precept a theory of organization can give: not a manual for dictators of any denomination more efficiently to subjugate human beings by the scientific application of Iron Laws, but a warning that the Leviathan of organization must not swallow the individual without sealing its own inevitable doom (L. von Bertalanffy, 1955: 82).

10.2 Bertalanffy's Organizational Work: The Formation of the Society for the Advancement of General Systems Theory and the Future Organization of the CASBS

10.2.1 A Scientific Society Dedicated to the Development of the General Systemology

Within the seminar "General Systems" of the CASBS the idea of the creation of a scientific society dedicated to the development of general systemology, already discussed in late 1953 between Boulding and Bertalanffy, was taken up again. Although it was initially Bertalanffy's idea and although it was also one of the proposals provided in the research project he wrote in 1954 to apply to the CASBS, the seminar report attributed the suggestion to create such a society to Boulding (late 1954), first called *Society for the Advancement of General Systems Theory* (SAGST). A committee was formed to specify its purpose, to define its status and its organization, and to study the conditions of its formation. And Bertalanffy was appointed as its executive secretary.

A first meeting dedicated to discuss these problems and to make this society known officially—"to sound out the opinion of those interested as to the most useful form of organization" (Hammond, 2003: 247)—was organized during the annual convention of the *American Association for the Advancement of Sciences* (AAAS), which was held in Berkeley from 26 to 31 December, 1954. This meeting took place within the framework of the division of philosophy of sciences of the AAAS.

His announcement in the programme of the convention looked like a manifesto. The announced general purpose of the SAGST was "to encourage the development of theoretical systems which are applicable to more than one of the traditional departments of knowledge" and its fundamental objectives were summarized in four points: "(1) to investigate the isomorphy of concepts, laws, and models in various fields, and to help in useful transfers from one field to another; (2) to encourage the development of adequate theoretical models in areas which lack them; (3) to eliminate the duplication of theoretical efforts in different fields; (4) to promote the unity of science through improving the communication among specialists" (L. von Bertalanffy, 1955: 74).

During this meeting Bertalanffy and Boulding presented the programme of the society they planned to create. The support was more substantial and enthusiastic than anticipated: between 50 and 80 people attended the meeting and thirty-three of them joined the society straightaway.[5] It was Bertalanffy's wife who was first entrusted with the task of registering, handling and even promoting these memberships (M. von Bertalanffy, 1973: 47). A modest funding provided by

---

5. Letter from R. Meier to L. von Bertalanffy (28/01/1958), BCSSS archive. Hammond, 2003: 247, 249.

a grant of the Bostrum Foundation helped to cover the first costs of the SAGST During this first meeting it was nevertheless agreed upon that its formal inauguration as an autonomous society was premature and that it should provisionally remain a committee of the division of philosophy of sciences of the AAAS In addition, the idea of a yearbook published by the SAGST, a compendium of articles coming from various academic fields and relevant to "general systems theory", was mentioned. Finally, an appointment for another meeting of the SAGST was made for the next annual congress of the AAAS (Hammond, 2003: 247). Meanwhile Bertalanffy started working at the CASBS on the legal details of the charter of the society. And the sessions of the seminar served as meetings of the SAGST

10.2.2 Bertalanffy's Contribution to the Reflection on the Organization of the CASBS

Besides the organizational work and his research together with Naroll, Bertalanffy worked on a number of projects relating to the development of the CASBS itself at the beginning of his stay in Stanford. He thus submitted two reports.

The first, entitled "On Interdisciplinary Exchange", concerned the manner how the CASBS could best be organized. In this report Bertalanffy first advocates "comprehensive presentations" in academic fields that are important for the study of the behavior. And by way of illustration he devises a complete programme that shows how biology could be expounded to the CASBS in a way to contribute to an "interdisciplinary integration". This programme organizes the discussion of all biological questions relevant to the study of behavior, in a way that responded to one of the ambitions of all principle founders of the centre: to base this study on biology, which serves consequently as a fundamental science. Bertalanffy advocates also, in a continuation of his last articles,

Chapter 10: The First Emigration to the US...

a second approach of the interdisciplinary integration with the creation of an "education for scientific generalists" focussed on "system problems". He presents this education as an answer to one of the fundamental problems that had justified the foundation of the CASBS: the growing complexity of modern science and society.

Finally in his second report, Bertalanffy, who then thought maybe of his uncertain future at the end of his stay in Stanford, recommended solutions to remedy a probably adverse effect of the annual rotation of the members of the CASBS: its orientation towards short-term results. In order to ensure the "continuity" of its activities, he suggests in particular that some members should be appointed for longer periods than one year[6]...

---

6. Two preliminary reports by L. von Bertalanffy to the CASBS (17/12/1954), BCSSS archive.

CHAPTER 11
Three Years in Los Angeles and Two Trips to Europe (1955-1958)

Bertalanffy's future was actually ensured at least in the short term by one of the other members invited to the CASBS His period of activity at the centre ended symbolically in September 1955 through a new meeting dedicated to the "theory of general systems", which took place this time in San Francisco and within the framework of the annual congress of the American Psychological Association. Bertalanffy gave a talk there on the principle of equifinality, Naroll on the principle of allometry in biological and social sciences, and the psychiatrist Alexander discussed his recent work on the principle of homeostasis in physiology and psychology (Hammond, 2003: 248). It was indeed Alexander who offered Bertalanffy an opportunity to secure his livelihood from November 1955 onwards.

11.1 Working next to Franz Alexander
11.1.1 Bertalanffy's Living and Working Conditions in Los Angeles

At that time Alexander created an "Institute for Psychosomatic Research" at Mount Sinai Hospital, Los Angeles. And he suggested to Bertalanffy to become its co-director. The latter was supposed to take care of the "somatic side" (i.e. being the director of biological research), while Alexander was supposed to be in charge of the "psychical side".[1]

---

1 Letter from L. von Bertalanffy to W. Westphal

Bertalanffy also found ways to stay in touch with teaching and to receive some additional income. He gave guest lectures on a voluntary basis at the Department of Physiology of the Medical School at the University of Southern California, where he met his friends Karl and Charlotte Bühler again, who had been assistant professors of psychiatry in this school since 1948. Sporadically he also gave lectures at the psychiatric clinic of Beverly Hills, whose founder was a Viennese expatriate, who had left at the *Anschluss* in 1938 and whom he had known little before that: Friedrich Hacker (1914-1989), whom he considered "a man of exceptional format".[2] His stay in Los Angeles was also interrupted by talks at several conferences: within the framework of the *Western Psychological Association* in Monterey and within the framework of the *Northern California Psychiatric and Psychosomatic Society* in San Francisco (Davidson, 1983: 63).

At the age of 54, Bertalanffy seemed thus to regain a stability he had lost almost ten years earlier. His job prospects were good and his salary was excellent. As for his son, he had completed his doctorate in Canada and had obtained a post as professor of histology at the University of Manitoba. Under these conditions, Bertalanffy and his wife bought a nice house in Spanish style in Beverly Hills, five kilometers from the hospital. And thus a happy period began,[3] whose sweetness was only matched by its shortness. Regarding Bertalanffy's research in Los Angeles, it was oriented towards three directions.

(18/11/1955), BCSSS archive. Davidson, 1983: 62; Hofer, 1996: 40; Brauckmann, 1997: 13-14.
2 Letter from F. Hacker to L. von Bertalanffy (28/10/1957), BCSSS archive. Davidson, 1983: 62; Hofer, 1996: 40-41; Brauckmann, 1997: 13-14.
3. Letter from L. von Bertalanffy to W. Wetsphal (18/11/1955); letter from L. von Bertalanffy to E. Rothacker (08/11/1956), BCSSS archive. Davidson, 1983: 64.

11.1.2 Bertalanffy's Commitment to Psychiatry: Studies on the Relations between Hallucinatory Drugs and Phenomena of Psychosis

Bertalanffy undertook first of all a critical study of significant re-orientation in the course of the psychiatric theory and practice, linked to their recent usage of psychotropic drugs (LSD and mescaline) in the study of psychosis. While the mental illness was until then seen as essentially psychogenic, its biochemical basis and new psychopharmacological therapies were henceforth at the centre of interest. And a new trend in psychiatry tended to consider the states induced by psychotropic drugs as equivalent to models of psychosis in general, giving birth to a series of physiological theories of this phenomenon. The study of schizophrenia in particular was then largely focussed on the "psychosomimetic" properties of these drugs.

Through a similar approach to that used in 1928 in relation to the problem of embryogenesis, Bertalanffy examined critically all current theories in order to show their limits and to highlight the points gained. He demonstrated the impossibility of reducing the study of psychosis to that of psychedelic states. This demonstration involved taking into account recent experiments that in his eyes fully confirmed the relevance of his principle of "primary activity" of the organism for psychiatry: similar hallucinations to those induced by psychotropic drugs appeared under conditions in which the subject was deprived of all stimuli. And Bertalanffy showed that despite their interest, all theories of psychosis, in particular that of schizophrenia, tended to neglect or even to ignore an essential dimension of such pathologies: their intimate relationship with symbolic activity. A psychosis can have its origin in a perturbation of the physiological functions, but it also reflects a perturbation of the symbolic functions and, as such, depends on the culture to which the subject belongs. As for the comparison of "peak experiences" (a term coined by Maslow), such as the mystical experience, with psychedelic states or schizophrenia, Bertalanffy considered it on the basis of reflections of his friend Huxley, as articulated in his *The Doors of Perception* (1954), and of a reinterpretation of his own studies on mysticism carried out in the 1920s. He showed the impossibility of such a comparison: even though both are cases in which the border between ego and non-ego is erased, the first is to be seen in a consistent and integrated symbolic universe which is absent in the second—thus explaining the fertility of the first and the sterility of the second.[4]

11.1.3 Some Developments of "Organismic" Psychology: Bertalanffy's Connection to Jean Piaget

The orientation of Bertalanffy's work towards psychiatry was in fact supportive of his orientation towards a psychology based on a transfer of his "organismic" concepts and principles to the study of mental life. This "organismic" psychology had already been outlined in 1951 on the occasion of his exchanges with Krech. In Los Angeles Bertalanffy brought in some developments which appeared in an article published in 1956, in which he expounded the problem of growth in its physical and mental aspects.

In this article Bertalanffy for the first time transposes his criticism of cybernetics to psychology. He tries to demonstrate that the concepts of "primary activity" and of development through differentiation are concepts equally fundamental on the mental as on the somatic level. And on this basis he attacks the ubiquity of the

---

4. L. von Bertalanffy, 1957, BCSSS archive. Letter from L. von Bertalanffy to A. Huxley (13/01/1954), in Gray and Rizzo, 1973: 202-203. Davidson, 1983: 63.

Chapter 11: Three Years in Los Angeles...

concept of homeostasis in contemporary psychology. Although it replaced the behaviorist pattern by introducing the concept of feedback, the homeostatic model of mental life remained for him fundamentally "mechanistic", as it explains it in terms of pre-established structures. In addition, this model appeared to him as essentially utilitarian: it comprehends mental events in terms of a logic of maintaining an equilibrium by neglecting the non-utilitarian aspects of mental activity, which appear in play, curiosity, or creativity. Bertalanffy thought that this model, beyond its epistemological inadequacy, presented a potential sociopolitical danger: that of turning the well-adjusted individual into the ultimate aim of education and that of generating well-oiled robots that maintain themselves in an optimal biological, psychological and social homeostasis (L. von Bertalanffy, 1956b).

There was a significant influence of Karl and Charlotte Bühler on this development of Bertalanffy, which, however, also had another important source: at that time Bertalanffy discovered a major justification for his transfer of the "organismic" principle of the primacy of gradual differentiation into psychology—i.e., in the final analysis, of dynamic prioritization (on the structural level)—in the work of Jean Piaget (1896-1980). In fact, Piaget's "genetic psychology" is focussed on the idea of a gradual formation of categories in the course of the mental development of the child. This formation corresponds to a process of differentiation based on a homogeneous set of undefined concepts and is founded on the child's interaction with its environment, more precisely on the "operations" that it carries out within it. Moreover, in this psychology Bertalanffy also found new arguments to support his perspectivism, insofar as it is based on a biological and cultural historicizing of categories.

11.1.4 Successful Research on Cancer Diagnosis

Despite this re-orientation towards psychology and psychiatry, Bertalanffy's main work in his Department of Biology related to the continuation of his research programme concerning the physiology of malignant growth and cancer diagnosis, a programme which he had not been able to complete in Ottawa. He managed to obtain grants from the *American Cancer Society* and the assistance of two very competent researchers, Francis and Marianne Masin.

Laboratory work started on July 2, 1956. Very quickly the last details of the technique of fluorescence microscopy, based on the use of acridine orange, were developed and hundreds of tests were carried out. The founding article by Bertalanffy and his assistants was sent to the journal *Science* on September 12 and was published on November 23.[5] Joined by Leo Kaplan, a doctor at the hospital, they applied this technique to the detection of gynaecological cancers in early 1957.

This diagnostic method was subjected to some criticism, which opposed to it the significant number of cases in which it is unreliable and in which an examination of the morphological properties of suspicious cells is necessary. This critique was unjustified, since Bertalanffy had himself straightaway admitted this necessity (M. von Bertalanffy, 1973: 47; Brauckmann, 1997: 14). Thus the publication of the article on the diagnosis of gynaecological cancers was first refused in June.[6] Nevertheless, he managed to get it published in another journal, and the method in question was soon used throughout

---

5. M. von Bertalanffy, 1973: 47; Davidson, 1983: 62. Letter from L. von Bertalanffy to H.M. Weaver (*American Cancer Society*) (19/05/1959), BCSSS archive.
6. Letter from the journal *Cancer* to L. von Bertalanffy (27/06/1957), BCSSS archive.

the world, not only for clinical studies of cancer, but also for basic research in embryology and virology: while Felix, Bertlanffy's son, took responsibility for demonstrating its effectiveness on several thousands of patients in his Canadian laboratory, the "Bertalanffy method" was praised and applied by scientists and doctors of big institutions, including, among others, the *National Cancer Cytology Center* in New York and Miami, the Pirovano Hospital in Buenos Aires, and the Institute of Molecular Biology of the Academy of Sciences of the USSR in Moscow.[7]

11.2 A First Trip to Europe and Lost Opportunities to Return There for Good

11.2.1 The Prospect (without Follow-up) of a Professorship in Hamburg and the First Return to Europe

An unexpected opportunity presented itself to Bertalanffy in July 1956, when he received a letter from his Viennese colleague in biology, Alfred Locker (1922-2005), announcing to him that the University of Hamburg created a chair for "theoretical biology and experimental medicine"[8] which corresponded perfectly with his profile—as Hamburg enjoyed a tradition in theoretical biology that had been established by Uexküll in the 1930s and 1940s.

This announcement came at an ideal time for Bertalanffy, who was just about to undertake a trip to Europe—the first since his emigration in 1949. He arrived

---

7. Letter from the journal *Cancer* to L. von Bertalanffy (27/06/1957); letter from L. von Bertalanffy to H.M. Weaver (*American Cancer Society*) (15/05/1959), BCSSS archive. M. von Bertalanffy, 1973: 48; Davidson, 1983: 63.
8. Letter from A. Locker to L. von Bertalanffy (02/07/1956), BCSSS archive.

on the European continent with his wife in September 1956. The couple stayed mainly in Switzerland and Germany, but spent also some days in Austria, Italy and France.

Bertalanffy's journey began in Geneva, where he had been invited to work from September 20 to 26 in a seminar on the "psycho-biological development of the child", organized within the framework of the World Health Organisation. On this occasion he met Lorenz (with whom he would afterwards remain in regular correspondence), the ethnologist Margaret Mead (1901-1978), the psychiatrist Erik Erikson (1902-1994), and above all Piaget. He contributed to this seminar also with a discussion of Piaget's theses, in which he contrasted his "organismic" conception of psycho-biological problems, based on the concepts of the "open system", of dynamic organization and "primary activity", with an, according to him inadequate, "cybernetic" conception, based on the concept of the "closed system", of "homeostatic equilibrium" and "primary reactivity" (L. von Bertalanffy, 1956b).

After this seminar he gave two other talks in Basle and Zurich before leaving for Germany. He came to Berlin on October 2 in order to give a talk on "The Contributions of System Theory to Contemporary Scientific Thought", which was broadcast on the radio. He also gave a talk on "New Paths in Biophysics" in Berlin on October 12 to the *Physikalische Gesellschaft*, as well as others in Heidelberg and Kiel.

In Kiel Bertalanffy talked with Kühnau, a biologist from Hamburg who had listened to his talk to discuss the project of an Institute of Theoretical Biology. But the modalities for its implementation were then not precise enough so that Bertalanffy could do nothing but to wait and see.[9] He would learn several months later that the

---

9. Letter from L. von Bertalanffy to A. Locker (08/01/1957), BCSSS archive.

11.2.2 The Two Lost Opportunities to Obtain a Professorship in Munich and Berlin

When he came back to Los Angeles with his wife during the first week of November, Bertalanffy discovered two letters from Rothacker with two new promises of a post in Europe. In the first, dated October 12, Rothacker responded to a letter from Bertalanffy, who had once again told him of his desire to obtain a chair in Germany, by informing him of the departure of Wenzl in Munich: in consequence of this departure, the head of the Faculty of Philosophy (Alois Dempf, 1891-1982) searched for a replacement for the chair of "Natural Philosophy". And Rothacker announced to Bertalanffy not only that he had personally recommended him for this post, but that Dempf had shown enthusiasm for this idea. He informed him also of the vacancy of another chair of "Natural Philosophy" in Berlin (that of Eduard May) and of the fact that he had personally recommended him, there again, to the head of the Faculty of Philosophy (Michael Landmann). In both cases Rothacker urged Bertalanffy to provide copies of his works.[11] And he wrote to him on October 23 to insist on the urgency of setting up his application, especially for Munich[12]...

But Bertalanffy found these letters only on his return two weeks later. He hastened to respond to Rothacker on November 8 by thanking him for his interventions,

---

10. Letter from L. von Bertalanffy to W. Wagner (27/07/1958), BCSSS archive.
11. Letter from E. Rothacker to L. von Bertalanffy (12/10/1956), BCSSS archive.
12. Letter from E. Rothacker to L. von Bertalanffy (23/10/1956), BCSSS archive.

by affirming to be honoured by the proposals in question and by attaching the required documents. He stated, however, that he did not wish to come back to Germany under any conditions, insofar as his position in Los Angeles guaranteed him great liberty in his research and a high living standard.[13] But it was too late (and perhaps tactless)... Rothacker answered him on December 18 to tell him that his application did not arrive in Munich in time and that Carl Friedrich von Weizsäcker (1912-2005) had decided to give preference to Wolfgang Stegmüller (1923-1991).[14] In fact Rothacker himself had finally recommended the latter. As regards the chair in Berlin, it had equally slipped out of Bertalanffy's hands because he had not provided the required documents in due course (Brauckmann, 1997: 13).

11.3 The Foundation of the Society for General Systems Research

Ten days after having received this distressing news, Bertalanffy went to Atlanta to attend the annual meeting of the SAGST, whose projects and first achievements gave a more positive note to the end of this year.

The first yearbook, which had been discussed two years earlier in Berkeley, entitled *General Systems*, had just been published thanks to the efforts of Rapoport and Bertalanffy. It started an series of 28 volumes. Rapoport and Bertalanffy would be its editors until the death of the latter. In addition to the yearbooks, the first issue of the monthly journal *Behavioral Science*, whose chief editor was Miller, was published in January 1956 (Hammond, 2003: 177). These two publications would very

---

13. Letter from L. von Bertalanffy to E. Rothacker (08/11/1956), BCSSS archive.
14. Letter from E. Rothacker to L. von Bertalanffy (28/12/1956), BCSSS archive.

Chapter 11: Three Years in Los Angeles...

quickly constitute the two major means of expression of systemic thought. In addition to those of Bertalanffy, Rapoport, Boulding and Gerard, the *General Systems Yearbooks* welcomed the contributions of authors such as Lorenz, Rashevsky, and Simon, but also those of the great cybernetician William R. Ashby (1903-1972), of Russell L. Ackoff (1919-), the founder of the "operational research", of Robert Rosen (1934-1998), the father of "relational biology", or even of the physicist and philosopher Mario Bunge (1919-).

Moreover, the SAGST was officially established, and its statutes and functioning were defined. It was nevertheless renamed *Society for General Systems Research* (SGSR), since its original name had unsuitably suggested the existence of a theory of general systems that needed to be promoted. As for its editorial and administrative office, it was established at the *Mental Health Research Institute* (MHRI) in Ann Arbor, Michigan, where not only Miller, who directed it after having worked at its foundation in 1954 and 1955, but also Gerard, Rapoport, and Boulding worked. The first election within the SGSR took place in May 1957. Boulding and Bertalanffy were elected president and vice-president respectively, and the psychiatrist Richard Meier, one of Miller's colleagues at the MHRI, was elected secretary and treasurer.

As Bertalanffy worked far away from Michigan however, he quickly played only a modest role in the concrete work of organizing and promoting the SGSR (above all, in the preparation of the annual meeting) on the one hand and in the edition of the *yearbooks* on the other hand, after having done the main work in the foundation during the first two years: Boulding, Meier, and Rapoport primarily took care of all this (Hammond, 2003: 154, 148-49). And this work was effective, since the number of members of the SGSR quadrupled between 1955 and 1957 (increasing from 33 to 147) and tripled between 1957 and 1958 (increasing to 494). In fact, *Behavioral Science* was used to promote the society, which naturally benefited from the success of its circulation (in late 1956 *Behavioral Science* already had a circulation of three thousand).[15]

11.4 The Conflicts with Alexander: A Second Voyage and the Search for a Position in Europe
11.4.1 The Strained Relationship between Bertalanffy and Alexander

Bertalanffy did not have the time to savor the at least temporary success of his ideas. Because even though he had left Ottawa, he was not for all that through with the relational problems in his work. Charlotte Bühler, who readily confided to him in 1956 that she admired him and wanted to work with him, told him in the same year about Alexander's disrespectful attitude towards him and towards her husband Karl.[16] And Bertalanffy himself soon took a profound dislike to Alexander, who to his taste made him feel his subordinate position a bit too much and, above all, tended to appropriate his work. The resentments and rivalries accumulated, and the year 1957 was particularly unpleasant for Bertalanffy; to the point that he finally described Alexander as the "archetypal Hungarian horse thief" and considered the two years at Mount Sinai Hospital as "the worst in my otherwise not really tranquil life".[17]

---

15. Letter from R. Meier to L. von Bertalanffy (28/01/1958), BCSSS archive. Hammond, 2003: 249.
16. Letters from C. Bühler to L. von Bertalanffy (20/05/1956 and 29/10/1956), BCSSS archive.
17. Letters from L. von Bertalanffy to F. Hacker (28/10/1957 and 27/12/1957), BCSSS archive. Brauckmann, 2003: 14. Translated from the German by ES.

Chapter 11: Three Years in Los Angeles…

11.4.2 Return to Europe: Fruitless Negotiations in View of a Chair in Gießen

A new trip to Europe in autumn 1957 gave him a breath of fresh air. He was in Rome in late October, and then, from November 4 to 9, in Geneva in order to work again within the framework of the WHO—this time in a "study group on ataractics[18] and hallucinogenics", where he outlined his work in this domain (L. von Bertalanffy, 1956b).

At the end of the year he had a new glimmer of hope of being able to leave America and to come back to Germany: the zoologist Wulf E. Ankel (1897-1983), Rector of the University of Gießen, offered him a chair of philosophy at his university on November 1, 1957, in order to unite the different branches of biology and medicine on the basis of teaching history and philosophy of science.[19] Back in Los Angeles, Bertalanffy replied three weeks later that the prospect of working once again for German science highly interested him, although his return was a difficult decision, given his substantial investment during the last two years in Los Angeles. He thus provided an agreement in principle, waiting nevertheless with his decision until he knew precisely the intellectual, material, and financial conditions of this post.[20] They were specified in January 1958 and satisfied Bertalanffy on the whole, who nevertheless demanded adjustments in the working conditions and especially a higher salary. Ankel granted him everything he wanted as regards the former, but made a new offer for his salary that was hardly higher than the initial offer.[21] Bertalanffy was, howev-

---

18. An ataractic drug is a tranquillizer.
19. Letter from W.E. Ankel to L. von Bertalanffy (01/11/1957), BCSSS archive.
20. Letter from L. von Bertalanffy to W.E. Ankel (25/11/1957), BCSSS archive.
21. Letters from W.E. Ankel to L. von Bertalanffy

er, motivated to go back to Germany, moreover to take a chair of "natural philosophy" which interested him such much that he "seriously thought about accepting the offer".[22] In March he simultaneously wrote to Rothacker and to Wilhelm Westphal (1882-1978), a biochemist in Berlin, in order to make sure that they supported him in case Ankel would try to obtain additional information about him from them.[23] But once again, this project did not succeed, the reasons for which remain unknown.

An important sign of his motivation to obtain a post in Germany at that time was his publication of no less than eight articles in German journals between 1956 and 1957, whose subjects covered the most important parts of his work: organic growth as well as the theory of open systems, his "organismic" approach to evolution and genetics, his philosophical anthropology and, of course, his general systemology. It is clear that Bertalanffy wanted to evoke the memory of the Germanic world.

11.5 A Profound Weariness of America and Repeated Efforts to Return to Europe

Bertalanffy's constant motivation to obtain a professorship in Germany had deeper causes than his conflicts with Alexander, even though they acted as a catalyst. The situation was similar to that in Ottawa, where his malaise had had causes beyond the conflicts with the university. This malaise exacerbated in fact throughout the years.

(10/01/1958 and 06/02/1958); letter from L. von Bertalanffy to W.E. Ankel (26/01/1958), BCSSS archive.
22. Letter from L. von Bertalanffy to W.E. Ankel (26/01/1958). Translated from the German by ES.
23. Letter from L. von Bertalanffy to E. Rothacker (13/03/1958); letter from L. von Bertalanffy to W. Westphal (13/03/1958), BCSSS archive.

11.5.1 Bertalanffy Disgusted by the American Relationship to Science

Bertalanffy described himself as "really tired of America". He considered it time to extract the essentials from his experience on the New Continent and could not bear the "American system" anymore, which did not let him do it.[24] "Ten years in America are enough" for him, not for scientific or financial reasons, but because he could not adapt to what he experienced as a commercialization of scientific activity and a loss of prestige for researchers, compared to their status in Europe. He compared the American scientist to "learned Greek slaves" who had been sold by "parvenus of the ancient Rome".[25] And, even though he acknowledged that the Americans had contributed significantly to two original and important developments of modern science, "behavioral science" and "interdisciplinary synthesis",[26] he did not support their general ignorance of European science and their propensity to proclaim ideas as new while they had long before been established in Europe.[27] He denounced the "well-known representatives of a class of the *systema naturae* which is perhaps interesting as regards the nature of their souls but does not stand out for its ethical principles":[28] those who appropriate the work of oth-

---

24. Letter from L. von Bertalanffy to E. Rothacker (21/04/1958), BCSSS archive. Translated from the German by ES.
25. Letter from L. von Bertalanffy to W. Wagner (28/06/1958), BCSSS archive. Translated from the German by ES.
26. Letter from L. von Bertalanffy to W.E. Ankel (25/11/1957); letter from L. von Bertalanffy to *Ford Foundation* (20/06/1958), BCSSS archive.
27. Letter from L. von Bertalanffy to R. Wagner (25/02/1958), BCSSS archive.
28. Letter from L. von Bertalanffy to R. Wagner (21/04/1958) Translated from the German by ES.

ers. He complained about being himself a victim of this phenomenon but believed it to be common, in particular among cyberneticians. All the more so as he, from 1958 onwards, maintained a frequent and dense correspondence with the physiologist Richard Wagner, one of Gessner's colleagues in Munich, whom he had contacted to protest against the fact that his name was never mentioned in cybernetics, while he had used the concepts of feedback and servomechanism in his works on muscle physiology from 1925 onwards. Henceforth Bertalanffy began to quote Wagner, as he thought that the Americans should know where cybernetics "actually" came from.[29]

11.5.2 New and Unsuccessful Attempts to Return to Germany or Austria—The Need to Leave Los Angeles and Menninger's Rescue

It was with Wagner that Bertalanffy discussed a possible new opportunity for him to get a position in Germany, after he had been informed by Gessner about plans for the foundation of an institute of biophysics in Munich and after Wagner had declared himself favourable to the idea of making this institute a "Max Planck Institute", directed by Bertalanffy. The latter considered such an institute "the best place that I could wish for".[30] Wagner and Gessner thus tried to organize his transfer to Munich, while Rothacker intervened once again to recommend him.[31] Wagner discussed it with Adolf F.J. Butenandt (1903-1995), Nobel Prize in Chemistry in 1939, who

---

29. Letter from R. Wagner to L. von Bertalanffy (03/02/1958); letters from L. von Bertalanffy to R. Wagner (25/02/1958 and 21/04/1958), BCSSS archive.
30. Letter from R. Wagner to L. von Bertalanffy (30/04/1958); letter from L. von Bertalanffy to R. Wagner (28/06/1958), BCSSS archive. Translated from the German by ES.
31. Letter from L. von Bertalanffy to E. Rothacker (21/04/1958), BCSSS archive.

also played an important role in the Max Planck Society. He was also just about to open an institute of bioenergetics in Munich, and although a certain Benzinger was already appointed for its direction, Wagner tried to convince his friend Otto Hahn (president of the Max Planck Society) and Butenandt to obtain the co-direction of this institute for Bertalanffy.[32] But these efforts would once again be in vain.

Bertalanffy had no illusions about this project and simultaneously searched for other opportunities to return to Germany or Austria. He thus envisaged creating an institute of cancer research in Germany, subsidized by the Ford Foundation or the Rockefeller Foundation in America.[33] But more importantly, he took up his project of creating an "Institute for Advanced Study in the Biological Sciences" in Vienna again and addressed the Ford Foundation in that respect in 1958. He learned in fact that the Ford Foundation planned the creation of an interdisciplinary research centre equally in Vienna and tried to convince the Foundation of the relevance of his own in this respect. He nevertheless received only a terse and very unkind reply that informed him of a categorical refusal to finance his project. He learned from this letter that Hayek had already proposed such a project in the spirit of the CASBS to this Foundation, and he hastened to write to him in March in order to ask him whether he could be integrated into it. But Hayek answered him that he had given up, as he had himself had very unpleasant experiences with the Ford Foundation.[34]

Unable to return to Europe, Bertalanffy had yet to find a means to leave Alexander by finding another job.

---

32. Letter from R. Wagner to L. von Bertalanffy (16/07/1958), BCSSS archive.
33. Letter from L. von Bertalanffy to R. Wagner (27/07/1958), BCSSS archive.
34. Letter from L. von Bertalanffy to F.A. von Hayek (19/03/1958); letter from F.A. von Hayek to L. von Bertalanffy (26/03/1958), BCSSS archive.

Moreover, this quest became increasingly urgent, as the Mount Sinai Hospital began to have financial difficulties in 1957 and as Bertalanffy thus had to abandon several research projects, especially in 1958. In addition to these, similar difficulties arose in Hacker's clinic, where he also taught at the same time (Davidson, 1983: 62; Brauckmann, 1997: 14). The solution came in the form of a psychiatrist whom he had got to know at the CASBS: Menninger.

CHAPTER 12
TWO YEARS IN KANSAS AND A THIRD VOYAGE TO EUROPE (1958-1960)

12.1 Bertalanffy's Activities at the Menninger Foundation

As early as October 28, 1957, Bertalanffy was in fact informed by Hacker that Menninger wanted to work with him.[1] Menninger had indeed followed his work during the preceding years, and he recognized his potential for psychiatry. He considered him ideally positioned to help him create, and later to direct, a department of research in biology,[2] which he wanted to establish within the Menninger foundation—a famous centre in the United States, founded in 1941 and dedicated to research and teaching in psychiatry as well as to psychoanalytic practice.

Things became more precise in December, when Menninger invited Bertalanffy and his wife to Topeka (Kansas) in order to discuss his project. The meeting took place in mid-January, 1958.[3] This collaboration and its terms suited Bertalanffy: the working conditions and the salary were excellent in his eyes, and he had the opportunity to pursue his studies that aimed at an integration of biology, psychiatry, and psychology into one "organis-

---

1. Letter from F. Hacker to L. von Bertalanffy (28/10/1957), BCSSS archive.
2. Letter from L. von Bertalanffy to W.W. Nowinski (27/05/1960), BCSSS archive.
3. Letter from L. von Bertalanffy to F. Hacker (27/12/1957), BCSSS archive.

mic" perspective. It was agreed upon and arranged before the month of June that he would work as guest professor at the Menninger Foundation from October 1, 1958, onwards—his status being nevertheless temporary.[4] His activity during his time in Topeka was divided in three directions.[5]

The first direction was that of teaching: during the first year in this institution he gave a series of lectures on the relations between biology and psychiatry. He was stimulated by the enthusiasm and the active participation of his listeners during these lectures, which helped him clarify and organize his ideas on this important topic in the perspective of his project of philosophical anthropology.

Bertalanffy also enjoyed excellent working conditions, which allowed him to write and publish twenty articles during his time there, most of them concerning his work on cancer. He was particularly interested in pulmonary cancer and wrote his articles in close collaboration with his own son, who had become a specialist in pulmonary cytology. These publications (ten publications between 1959 and 1960 and four in 1961) responded to some extent to the need to justify the subsidies the *American Cancer Society* had granted him. But the most remarkable text published by Bertalanffy during his stay in Kansas was without doubt "The Psychopathology of Scientism", which constituted a formal attack on positivism not only in terms of the theory of knowledge but even more concerning its ideological and social consequences: based on Hayek's analyses, he tried to demonstrate that positivism was ultimately the "epistemological root of scientism", and that the latter led logically to a conscious and totalitarian planning of social

---

4. Letters from L. von Bertalanffy to R. Wagner (28/06/1958 and 23/09/1958), BCSSS archive.
5. Letter from L. von Bertalanffy to K.A. Menninger (01/07/1959), BCSSS archive.

phenomena, which was not a prophecy but a contemporary reality incarnated in "human engineering" and mass persuasion. Bertalanffy linked scientism closely to a diagnosis of sick Western societies: he considered mental illness and delinquency, whose increase was striking in the United States of the 1950s, as the price to pay for a, scientifically guaranteed, greater comfort and conformity of an individual who was denied in his or her essence (an "immanent activity"; L. von Bertalanffy, 1960: 214) by a mass society that established utilitarianism and mercantilism as supreme virtues, and conditioning as a means of accomplishing its logic. For him it was not stress but an "inner emptiness" and an "outer enforcement" (L. von Bertalanffy, 1960: 214) that combined to create the conditions for these mental and social disorders. Perspectivism, in that it is opposed to this "epistemological root of scientism" (L. von Bertalanffy, 1960: 202-204), i.e. posititivism, thus appeared to Bertalanffy as the preliminary and fundamental step that paved the way for a redefinition of the relations between science and society in a less fatal direction.

In addition to his teaching and research, Bertalanffy worked on plans and activities of the new biology laboratories to be established at the Menninger Foundation. He developed detailed research programmes, oriented in particular toward the histochemistry of the central nervous system. First, the idea was to establish these laboratories in such a way that they were ready for use on June 30, 1959, the day when Bertalanffy's status as guest professor ended. Bertalanffy also intended to dedicate one part of his future research department to the pursuit of his work on cancer, which would permit him to extend his substantial funding. But two factors were opposed to these plans.

12.2 Bertalanffy's Difficulties at the Menninger Foundation

The first factor was related to the fact that the institutions that supported the Menninger Foundation, believing that it had an essentially psychiatric and psychoanalytic purpose, were reluctant to subsidize biology laboratories there, especially in the directions of research that Bertalanffy wanted to give them, despite that fact that Menninger himself was very much in favour of it.[6] Until early 1960 Menninger and Bertalanffy did everything in their power in order to convince these institutions,[7] but it was in vain, and the whole project was abandoned before June.[8]

The second factor was due to "manoeuvres" carried out at the Mount Sinai Hospital in Los Angeles after Bertalanffy's departure. Kaplan, who had collaborated in his research programme on cancer, was now director of the biology laboratories. And he managed to transfer the subsidies previously granted to Bertalanffy by the *American Cancer Society* in his direction in May 1959, finding support for minimizing Bertalanffy's contribution to the research on the diagnostic technique of fluorescent microscopy with acridine orange.[9] In addition, a conflict during the same period matched Bertalanffy directly against Kaplan, insofar as the latter wanted to publish an article on this technique by putting Bertalanffy's name in the final position on the list of coauthors, after his own. Bertalanffy protested vigorously, arguing that he had di-

---

6. Letter from L. von Bertalanffy to W.W. Nowinski (27/05/1960), BCSSS archive.
7. Letters from L. von Bertalanffy to K.A. Menninger (27/01/1960 and 16/02/1960), BCSSS archive.
8. Letter from L. von Bertalanffy to J.M. Scher (24/06/1960), BCSSS archive.
9. Letter from L. von Bertalanffy to H. Weaver (*American Cancer Society*) (19/05/1959), BCSSS archive.

rected all this research and stood at its origin so that his name should thus be listed at the top of such a list—but to no effect. "Fed up with this affair" and with a "feeling of deep embitterment" he finally abandoned an obviously futile struggle against this appropriation of his own work.[10] This experience left an all the more bitter taste in his mouth as he kept receiving, not only from the United States but also from Germany in 1959 and 1960, confirmations of the effectiveness of the diagnostic method he had developed, henceforth tested on tens of thousands of cases.[11]

12.3 The Repeated Search for a Position in Europe

Bertalanffy, for whom Menninger had managed to obtain a provisional extension of the status as visiting professor for the year 1959/1960, therefore needed to find a new academic position.

On May 27, 1960, he thus wrote to the biochemist Nowinski, an old friend of his, to inform him of his availability for a possible post at the university where the latter worked (in Texas).[12] But there was no reply to this letter.

The plan to create a Max Planck Institute of biophysics or "theoretical biology" in Germany, of which he would be in charge, was also put on the agenda. Sup-

---

10. Letter from L. Kaplan to L. von Bertalanffy (20/05/1959); letter from L. von Bertalanffy to L. Kaplan (02/06/1959); letter from L. von Bertalanffy to H. Weaver (*American Cancer Society*) (19/05/1959), BCSSS archive.
11. Letter from L. von Bertalanffy to N. Schümmelfelder (08/01/1959); letter from W. Sandritter to L. von Bertalanffy (05/04/1960); letter from L. von Bertalanffy to H. Weaver (*American Cancer Society*) (19/05/1959), BCSSS archive.
12. Letter from L. von Bertalanffy to W.W. Nowinski (27/05/1960), BCSSS archive.

ported by Gessner, Bertalanffy wrote to Butenandt on January 19, 1960, in order to win his support for this project,[13] considering his status as a Nobel Prize winner and his influence in the Max Planck Society. Even if the idea for such an institute was again not followed up, Butenandt nevertheless joined Wagner to support Bertalanffy's candidacy for the direction of a laboratory of research on cancer, which was advertised at the Faculty of Medicine in Hamburg. Butenandt even wrote a letter to Helmut Coing, the chairman of the *Deutsche Wissenschaftsrat* (German Council of Science and Humanities), in order to win his support.[14] And he wrote to Bertalanffy in July to arrange a meeting in Munich on October 6 in order to meet him personally.[15]

Bertalanffy had indeed planned a trip to Europe, where he went from August to late October 1960, after having notably participated in an important conference on the "principles of self-organization" at the University of Illinois in June—alongside Ashby, Hayek, Rapoport, Heinz von Foerster (1911-2002), and Warren MacCullough (1898-1969). During his trip to Europe, Bertalanffy spoke at several conferences in Italy and Germany: he expounded his general systemology in Florence, and his "organismic" perspective in biology and medicine in Regensburg. After having met, as previously arranged, Butenandt in Munich, he went to Hamburg and talked with the dean of the Faculty of Medicine on October 23 in order to discuss the project of a laboratory of research on cancer. Unfortunately for him, a clash of interests within the university hospital ruined this project: the physicians and surgeons opposed the creation of such a laboratory for biologists (Brauckmann, 1997: 15).

---

13. Letter from L. von Bertalanffy to A. Butenandt (19/01/1960), BCSSS archive.
14. Letter from R. Wagner to L. von Bertalanffy (11/07/1960), BCSSS archive.
15. Letter from L. von Bertalanffy to R. Wagner (23/07/1960), BCSSS archive.

Chapter 12: Two Years in Kansas...

Thus, on his return to America in late October 1960, at the age of 59, Bertalanffy found himself unemployed. Before leaving for Europe he had received a proposal by the psychologist Ralph Gunter, who had invited him to work with him in Northridge. There had even been negotiations on the subject. But Bertalanffy had eventually declined the offer, considering the salary too low and the teaching hours (12 hours per week) too high for being able to devote himself to writing the various books he had planned.[16]

---

16. Letter from L. von Bertalanffy to R. Gunter (11/07/1960), BCSSS archive.

CHAPTER 13
The Second Canadian Period: Edmonton and Frequent Trips to Europe (1961-1969)

Bertalanffy's salvation came from a psychologist he had met in Los Angeles and who had felt great admiration for him since then: Joseph R. Royce. In 1961 he invited him to come back to Canada, to the University of Alberta, Edmonton, where he directed the Department of Psychology, so as to help him create a "center for advanced studies in theoretical psychology". In order to assure him a comfortable situation, Royce allowed Bertalanffy to obtain an additional post as professor of theoretical biology in the Department of Biology at the same university (Davidson, 1983: 65).

13.1 Bertalanffy's Liberty and Deployment in Edmonton

The offer was a boon for Bertalanffy, who was happy to find himself finally, at the age of 60, "professor 'at large' without definite teaching obligations but with connections with several departments", and with the possibility to profit from this liberty in order to bring into form the scattered notes he had accumulated for years.[1] If he affirmed several months after his installation to be "fully satisfied" with the people he met and with his environment,[2] he would not have to modify his judg-

---
1. Letter from L. von Bertalanffy to A. Huxley (31/10/1961), in Gray and Rizzo, 1973: 209.
2. Letter from L. von Bertalanffy to K. Lorenz (22/03/1962), BCSSS archive.

ment later, despite the fact that his judgment regarding Canada in general did not change—namely, that it was a "spiritual desert", where it was "no wonder that one became arthritic and suicidal".[3]

13.1.1 Bertalanffy's Administrative Activities between 1961 and 1963

During his first two years in Edmonton Bertalanffy put a lot of energy particularly into the development and the finalization of two projects.

The first was carried out in collaboration with Maslow and the sociologist Pitirim A. Sorokin (1889-1968) and led to the creation of the "International Center for Integrative Studies" in New York in 1962 (Brauckmann, 1997: 16).

Together with Royce, the philosopher Hermann Tennessen, and the psychiatrist T.E. Weskowicz, he worked on the second project, namely the establishment of a *Center for Advanced Study in Theoretical Psychology,* which was the reason for his coming to Edmonton and which became one of the world's most important institutions dedicated to psychology throughout the three following decades. The definitive programme of the centre was formulated in November 1963, and the centre began to operate in 1964. Its goals were similar to those Bertalanffy assigned to theoretical biology in 1932: to create an inventory and a critical evaluation of already formulated psychological theories in order to go beyond controversies and to retain that which could be considered valid; to formulate new theoretical points of view capable of establishing order in the mass of empirical data that was already available; and to work on the development of a philosophy of science in general and of psychology in

---

3. Letter from L. von Bertalanffy to A. von Hochstetter (16/09/1965), BCSSS archive. Translated from the German by ES.

particular, capable of overcoming a "physicalistic-mechanistic" vision that was still largely dominant.[4] From the moment that the centre was established, Bertalanffy gave regular talks on the history and the theory of biology there (Brauckmann, 1997: 17).

13.1.2 A Period of Synthesis and Intellectual Liberty

Bertalanffy had so much liberty in Edmonton that he managed to find the time to give free rein to his secret passion by publishing four articles on the history of stamps (in 1962, 1963, 1967, and 1969) in Italy and by publishing a book on this history in 1968, with which he revealed himself to be one of the greatest specialists of the time. But above all he turned his new situation to good account so as to deepen and synthesize his reflections in four directions, which formed many additional occasions for a development of philosophical anthropology, for which he had outlined plans since 1947. The main result of it was his work *Robots, Men and Minds*, published in 1967.

The first direction was a systematization of his philosophy of symbolism, which nevertheless did not present any new conceptual element compared to the scattered reflections published on the subject in various articles since the Alpbach meeting. This philosophy was the subject of his talk at a conference on "psychology and the symbol" in 1962, organized by Royce in San Francisco, and provided him with the contents of a long essay published in 1965 in the conference proceedings.

Bertalanffy also exposed the epistemological consequences of his perspectivism in several articles (including one published in 1965 on the "History of Theoretical

---

4. Letter from J.R. Royce, L. von Bertalanffy, H. Tennessen and T.E. Weskowicz to H.S. Armstrong (vice-president of the University of Edmonton) (18/11/1963), BCSSS archive.

Models in Biology", by explaining a philosophy of models and of modelling that had in fact operated in his works since the 1930s. While criticizing the recent "feverish 'model building'", frequently developed "as a purpose in itself and often without regard to empirical fact" (L von Bertalanffy, 1969: 100f.), he theorized the need for such conceptual constructions whose fictitious, free, and non-monopolistic nature was in his eyes the very reason for their effectiveness. They constitute the only means of a conceptual mastery of nature and should only be judged on pragmatic criteria, based on their explanatory and predictive merits. According to Bertalanffy all scientific theories are models in the broad sense of the term. His perspectivism finds its conclusion in the idea that "the model signifies the essence of all knowledge",[5] which is precisely the source both of the limits and the fertility of creative scientific thought.

A third direction adopted in Bertalanffy's work in Edmonton, which expresses itself in particular in an article that was published in 1964 and was entitled "The World of Science and the World of Values", was his revival of a "critique of culture" [Kulturkritik], which had marked his beginnings and found ample material for reactualization in the contemporary context. His "modelling" allowed him first of all, as he had already done in 1960 (L. von Bertalanffy, 1960: 210-211), to rehabilitate the tradition of a theoretical history that, according to him, developed from Giambattista Vico (1668-1744) and Arnold Toynbee (1889-1975) to Spengler: these authors had in his view provided "molar" and qualitative models that broke with classical "molecular" (or ideographic) approaches, which should not be judged *a priori* (in particular as regards moral considerations) but, as every model, concerning their explanatory and predictive value. And, as in every science, if such models should

---

5. L. von Bertalanffy, 1965: 298. Translated from the German by ES.

present a danger, it is not intrinsic to the model but due only to a neglect of their perspectivist nature, to the assimilation of the model to a metaphysical reality. This rehabilitation, which served as an extension of the general systemology to history, also supported a comeback of the Spenglerian "decline of the West" in Bertalanffy's publications. He believed that Western culture was exhausted and the twentieth century showed the emergence of a new and strange civilization: a "global, technological, mass civilization" (L. von Bertalanffy, 1964). In the wake of his criticism in this regard, initiated in 1960, he denounced a loss of values on all levels of culture, a "lost soul" induced by a materialist, utilitarian, and mercantile logic which constitutes a constant denial of the symbolic dimension of human existence and, ultimately, of the value of the individual. He particularly attacked the "degradation of the democratic dogma" in the domains of education and research into an egalitarianism, an environmentalism, and a frenzied "cult of the group", which contribute to producing a general mediocrity (L. von Bertalanffy, 1964). This "degradation" inherent in the advent of the "masses" is to the political order what loss of energy is to closed physical systems; hence the ideological dimension of the concept of the open system which, beyond science, subsumed for Bertalanffy all processes that thwart the destructive tendency of the mechanistic vision of the world. To a levelling down, to a society that tends to deny real differences, to a substitution of the legitimate quest for a social status with "empty and silly symbols", and to a confusion of education with "human engineering" Bertalanffy opposed the old ideal of *Bildung*, which suggested for him a non-utilitarian education and science, destined to permit a "full unfolding of human potentialities", and a production of individuals worthy of the name, who come to terms with themselves in a free society; but also the rehabilitation of a hierarchical society, where "status symbols must be replaced by status" and where "the empty chromium symbol of the

Cadillac can profitably be replaced by recognition of spiritual aristocracy" (L. von Bertalanffy, 1964).

In addition to these three directions Bertalanffy also worked on a reconsideration of the classical problem of the relations between body and spirit, in particular in the light of modern physics and modern developmental psychology. This reconsideration was expressed in an article published in 1964, in which Bertalanffy tries to show the inadequacy of the classical theories of parallelism, of interaction, and of psychophysical identity, departing from the "fact" that they are, according to him, ultimately rooted in an outdated dualist Cartesian problematic. He emphasizes that this dualism is not an essential and elementary fact but the result of a specific cultural and scientific evolution. In his eyes "matter" and "spirit" are reifications that are no longer relevant in modern science. And he argues that it is possible to extract the problem of the relation between mental life and its material substrate from the metaphysical framework in which it had remained until then to relate psychology and neurophysiology in an unprecedented and non-reductionist manner: namely by postulating an "isomorphism" between the conceptual constructions and both disciplines, which does not imply any similarity between the psychological and neurophysiological processes, but needs to be thought in terms of a common code—in the manner of the relation that exists between a manufacturing programme and the result of this programme. In other words, according to him the "psychophysical" unity needs to be understood in terms of generalized formal constructions and thus as "neutral" with regard to psychology and physics: as is the particular meaning of his general systemology.

The treatment of this problem of the relations between "body" and "spirit", however, was for Bertalanffy only an aspect of the project of systematizing his "organismic" psychology, which he carried out between 1964 and 1966. His wish to integrate biology, psychiatry, and psychology in the same conceptual framework was accomplished in Edmonton. And, more generally still, this systematization provided his general systemology not only with the extension to which he had devoted it programmatically since 1945, but also with an ethical dimension that had been only little explicit until then.

13.2 "Organismic" Psychology: General Systemology as the Matrix of a "Humanistic" Science

On January 13 and 14, 1966, Bertalanffy gave two lectures a Clark University, Worcester (MA), which marked in particular his opposition to Freudian psychoanalysis, which was then fashionable in the United States. In "Robots, men and minds: the organismic concept in psychology and biology" and "The open system of science: toward a new natural philosophy" he gave a broad outline of two books published in 1967 and 1968 respectively: *Robots, Men and Minds* summarizes all his work since his emigration to America and marks in fact the end of his intellectual career, and *Organismic Psychology and System Theory* is an extract from the previous publication limited to considerations relating to psychology.

Bertalanffy's initiation to psychology was an axiom of his philosophy of knowledge since its beginnings and is reminiscent of his involvement with neo-Kantian circles in Vienna and Germany: he conceived of cognition as an active process, opposed to "the dogma of immaculate perception" (L. von Bertalanffy, 1967: 91), which considers the organism as a passive receptor of stimuli. There is no perception of things "in itself"; perception is not a reflection: the organism creates the world that surrounds it on the basis of linguistic, symbolic, emotional, intentional and other factors. It is the same human symbolism which gives "consistency" to the perceived world and which organizes a mass of otherwise confused sensations.

The heart of his psychology is the transfer, from biology to psychology, of his model of the organism as an "open system" that is capable of staying away from the state of equilibrium. The "organismic" psychology which derived from it was in itself not an original contribution on his part. Bertalanffy joined *Gestalt* and humanistic traditions, represented by authors he had discovered after his departure from Austria: in particular Piaget, Maslow, Karl and Charlotte Bühler, Heinz Werner (1890-1964) and Gordon W. Allport (1897-1974); but also, from a more psychiatric point of view, Kurt Goldstein (1878-1965), whose works he had got to know in the mid 1930s. Like these traditions he rejected the "mechanistic" vision of personality, centred on the concepts of instinct, reaction and conditioning, and opposed to them those of creativity, development, individuation, learning, differentiation, emergence, self-realization and self-transcendence. Bertalanffy's specific contribution, acknowledged by the "humanistic" trend in psychology, consisted in the fact that he provided biological foundations to this psychology, which permitted it to develop and, more generally, established a continuity between the two disciplines; but also in the depth of his criticism against behaviorism, which was similar to his criticism of cybernetics.

His conception of man as an open system characterizes him by his primary, creative and intrinsically directed activity, while it insists on understanding him in his interaction with his social and cultural environment. It is also opposed to the "model of the robot", which reduces him to a passive automaton, entirely subject to the conditions of his environment and directed towards the maintenance of a homeostatic equilibrium. Bertalanffy rejected not only behaviorism, but also the Freudian vision of behavior controlled by the desire to eliminate tensions and to achieve stability: as an open system, man,

Chapter 13: The Second Canadian Period...

on the contrary, is far from such an equilibrium; this is precisely the source of his creativity and greatness. And even though the behaviorist pattern of "stimulus and response" refers to a psychological reality, it corresponds in fact according to him to secondary regulatory mechanisms that are superimposed on the primary activity of the organism and cannot characterize his behavior.

Indeed, for Bertalanffy the dominance of the "model of the robot" was, like that of a "mechanistic" biology, an expression of the "Zeitgeist of a highly mechanized society" (L. von Bertalanffy, 1967: 12; cf. L. von Bertalanffy, 1949: 169). And the model's systematic tendency to explain behaviour in utilitarian terms was the academic response to a "commercial" society whose supreme values were to maximize profits and to minimize economic costs. In his eyes the problem was that this model, by reifying man, legitimated, favoured and served as a basis of a "behavioral engineering" that opened the door to a manipulation and training of individuals, for which mass media and advertising were the main sites of implementation.

From the same perspective, Bertalanffy attacked the patterns of functionalist sociology (of Talcott Parsons (1902-1979) in particular). Although this field developed a systemic approach, it was of a "cybernetic" and not an "organismic" type, which, based on the concept of homeostatic equilibrium, entirely neglected social development and evolution and even more the fact that they rest on the creativity and the spontaneity of individuals.

In response to criticism that was more and more frequently addressed to a systemic movement accused of technocratism, Bertalanffy joined Lewis Mumford (1895-1990) and his critique of the "megamachine" (L. von Bertalanffy, 1969: 196) to reject any submission of the individual to "the Leviathan of organization" (L. von Bertalanffy, 1955: 82), in particular any idea of a scientific control over society, which largely inspired some of his fellow founders of the systemic movement, such as Miller and Gerard. During the 1960s in general and more particularly in 1967, he consequently also attacked a whole trend within the systemic movement which he considered a perversion characterized by systemic inspiration as he understood it, known as "systems analysis". Not only does this trend interpret symbolism simplistically in terms of information and in this way evacuate questions of meaning, but it contributes to technologies of social planning and of war and, hence, to the slavery of humanity. In contrast, Bertalanffy saw in his general systemology a "natural philosophy" (L. von Bertalanffy, 1967: 53f.), which can give rise to a "humanistic" (L. von Bertalanffy, 1967: 114-115) science, capable of supplying more appropriate models of human reality with an acute awareness of their perspectivist character, and to thereby found the very means of working towards a technology and a society that respect the needs and the essence of the individual. And beyond this general systemology he appealed to the intellectuals to assume their responsibilities: as "great spectacle makers of History" (L. von Bertalanffy, 1967: 52) they have the means to preserve "the remnants of old culture" (L. von Bertalanffy, 1967: 12) and to work towards a humanization of science instead of towards its technological exploitation in the service of the alienation of the individual.

13.3 Regular Trips to Europe

Bertalanffy's liberty in Edmonton consisted also in travelling: he increased the number of talks he gave at conferences, whether in Canada, the United States, or in Europe, to where he returned each year between 1964 and 1969.

His first trip to Europe took place in September and October 1964: he participated in several conferences in Germany and Austria, which related to Nicholas of Cusa, the history of civilizations, as well as to the biophysics of the open system and cancer research.[6]

The second journey took place mainly in Austria in July and August 1965. It was first and foremost a family holiday and did not include any conference.[7] But Bertalanffy met enough colleagues to find out that his ideas were known and influential in Europe.[8]

Bertalanffy also spent the summer of 1966 in Europe. He gave a series of lectures in Hamburg on the position of man in the modern world and another entitled "Symbolism and Anthropogenesis" at a conference in Bad Homburg. In Vienna he also met a childhood friend whom he had lost sight of since his schooldays: the composer Jelinek, for whom he arranged a concert in Canada during the following months.[9] In addition, Bertalanffy had the opportunity to ascertain that his place was not at the University of Vienna, where the best had left (such as Lorenz) or died (such as Reininger), and where in his eyes only the mediocrity "of a generation of fools" [Trottelgeneration], who preferred seeing him work in Canada instead seeing him come back to Austria, prevailed.[10]

However, he also had some admirers and "spiritual heirs" in Vienna, for instance, the physiologist E. Rohracher, who defended his ideas and whom he met in 1967, twenty years after having left Vienna, or other biologists, such as Locker and Rupert Riedl (1925-2005). His general judgment was nevertheless confirmed when

---

6. Letter from L. von Bertalanffy to K. Lorenz (19/03/1964), BCSSS archive.
7. Letters from L. von Bertalanffy to K. Lorenz (22/06/1965 and 04/10/1965), BCSSS archive.
8. Letter from L. von Bertalanffy to A. von Hochstetter (16/09/1965), BCSSS archive.
9. Correspondence between L. von Bertalanffy and H. Jelinek (21/03/1966-19/11/1966); letter from L. von Bertalanffy to M. Kiepach (17/11/1966), BCSSS archive.
10. Letters from R. Friedmann to L. von Bertalanffy and from L. von Bertalanffy to R. Friedmann (20/02/1967 and 03/03/1967), BCSSS archive.

in 1968, precisely through Locker and Riedl, he learned that rumors were going round in Vienna according to which he was a laboratory assistant in a Jewish hospital in Canada …[11]

Throughout the same journey to Europe in 1967, Bertalanffy also went to Eastern Germany at the invitation of the German Academy of Sciences Leopoldina (which elected him among its members two years later), in order to contribute to a conference on "model and knowledge" (October 19 to 22) with a talk about the distinction between the cybernetic model and the model of the "open system".[12] This invitation was in fact a sign of a more general interest his thought aroused in the Eastern European countries.

13.4 The Interest Bertalanffy Aroused in the Eastern European Countries

Bertalanffy's ideas were discussed in the USSR and the European countries under Soviet influence since the early 1960s. More generally there was a connection between American and Soviet "behavioral sciences" at that time, after a delegation of scientists headed by Miller and including in particular Gerard and Rapoport had come to Moscow in May 1961 at the request of President Eisenhower (Hammond, 2003: 179). This connection was part of the outcome of the publication of several articles that were written by Soviet system theorists and translated into English by Rapoport

---

11. Correspondence between L. von Bertalanffy and H. Rohracher (01/08/1967-09/05/1968); letter from F. Gessner to L. von Bertalanffy (23/10/1968); letter from L. von Bertalanffy to F. Gessner (07/11/1968); correspondence between L. von Bertalanffy and A. Locker, BCSSS archive.
12. Letters from K. Mothes to L. von Bertalanffy (21/06/1966 and 27/02/1968); letter from L. von Bertalanffy to K. Mothes (22/09/1967), BCSSS archive.

in *General Systems* in 1960. It soon led to the discovery in the West of *Tectology*, published in 1913 by Alexandr Bogdanov (1873-1928), who clearly anticipated Bertalanffy's general systemology by outlining a "General Science of Organization".

The interest in Bertalanffy's works was emphasized by different philosophers and scholars of the Eastern European countries who, although critical of what they regarded as his "positivist idealism" and insisting on what they considered the shortcomings of his thought compared to that of Engels, wanted to see in him the evidence of a progress of "dialectical materialism" in the West and in this materialism the true philosophical foundation of his general systemology.

In 1965 Bertalanffy began to correspond with several of them, notably with the Eastern-German philosopher Arno Bendmann, who worked on a systematic study of the philosophical implications of his "organismic" philosophy of life and typically aimed at demonstrating the congruence with "dialectical materialism". Bertalanffy objected to this ideological appropriation at the time. While acknowledging a certain parallelism, he denied any influence. And even though he admitted to the dialectical character of his philosophy of knowledge, the same was not true for his materialism; he objected particularly to the "sociologist" reductionism of Marxism.[13]

13.5 His Last Years in Edmonton

Apart from his travels, Bertalanffy dedicated the years 1966, 1967 and 1968 to writing and publishing several books: after *Robots, Men and Minds* in 1967, six books were published in 1968. Besides *Or-*

---

13. Letter from L. von Bertalanffy to I. Kanaev (29/11/1965); letters from L. von Bertalanffy to A. Bendmann (31/03/1966 and 12/04/1966); letter from L. von Bertalanffy to F. Gessner (18/10/1968), BCSSS archive.

ganismic Psychology and System Theory and his very remarkable *History of the Letter, from Early Times to 1840*, he published in particular a volume almost exclusively consisting of already published articles, most of which were English translations of articles originally written in German. Entitled *General System Theory: Foundations, Developments, Applications*, it would constitute Bertalanffy's most often published and best known book, despite its heterogeneous, non-chronological and unsystematic nature—a fact that has significantly contributed to misunderstandings about his thinking.

Having reached retirement age in September 1966, Bertalanffy sent a request for a one-year extension of his contract for the year 1967/1968 to the administration of his university. It was granted, as would be the same request in 1968 for the year 1968/1969. His problem was that his past life, especially the fact that he emigrated from Austria as late as 1948, left only little hope for even a ridiculous pension. This problem seemed to have found a solution when he applied for the annual prize of fifty million dollars awarded by the Royal Bank of Canada to reward scientific work in 1967: he enjoyed the support of three Nobel-prize winners—beside Butenandt, he was supported by Walter R. Hess (1881-1973), with whom he was also in correspondence,[14] and by the biologist Albert Szent-Györgyi (1893-1986). But despite this support, he did not receive the prize (Brauckmann, 1997: 17).

Bertalanffy returned to Europe in the summer of 1968. He thus came back to Alpbach twenty years later in order to participate, notably along with Piaget, Hayek, Weiss, Waddington and the psychoanalyst Viktor E. Frankl (1905-1997), in a conference entitled "Beyond Reductionism: New Perspectives in the Sciences of Life". This conference was co-organized by Arthur Koestler (1905-1983), who had devoted himself for several years

---

14. Letters from L. von Bertalanffy to W.R. Hess (16/02/1967 and 05/09/1967), BCSSS archive.

to biological philosophy and developed holistic conceptions that were largely and explicitly inspired by Bertalanffy's thought. Both men mutually appreciated and admired one another, and they would remain friends after this conference, where Bertalanffy expounded his "organismic" vision of phylogenetic evolution and his criticism of the synthetic neo-Darwinian theory in terms that had not changed since *Das biologische Weltbild*—this invariance was at least partially justified by the consistency of his aporias of this theory, which moreover led Bertalanffy to the works of an equally critical young researcher for whom the future would hold great fame: Stephen J. Gould.[15]

In 1969, Bertalanffy, who could no longer claim another extension of his status as professor at Edmonton, was offered the opportunity to finish his career in the United States at the State University of New York (SUNY) at Buffalo, where the official retirement age was seventy years. He accepted the post, which would be his last academic position.

---

15. BCSSS archive (offprints).

CHAPTER 14
A Last Return to the United States: Twilight Years in Buffalo (1969-1972)

Bertalanffy took up his post at SUNY on September 1, 1969. Like John C. Eccles, the Nobel-prize winner of 1963 who became his colleague, he was posted both at the Faculty of Social Sciences and at the Department of Theoretical Biology (where he also met Rosen again). His relationship with Eccles was relatively distant, insofar as his agnosticism could hardly put up with the religious overtones by which the thought of his colleague was marked (Brauckmann, 1997: 18).

14.1 Teaching at SUNY

Bertalanffy's teaching at Buffalo was limited to a seminar of three hours per week, where the relations between science, culture and society, or science and systems philosophy were discussed (Davidson, 1983: 66).

Although his course started at 4 p.m. on Fridays, he always had a capacity audience of never less than fifty listeners (Brauckmann, 1997: 18; Davidson, 1983: 66). His philosophy, which implies a radical critique of a world that engenders "the desperate search for new values and ways in the discontent of our present technological, mechanistic, behavioristic, industrial and military civilization", at that time attracted a "counter-culture", which very often saw in one of his best friends (Huxley) a spiritual father, all the more as Bertalanffy warned against this phenomenon: he did not hesitate to express not only his contempt for a "society of affluent mediocrity", which inundated the world with its vulgarity, and for a democracy reduced to "a herd of remote controlled cattle", but also for hippies, "higher consciousness", pseudo-alternative political movements, "American yoga", Zen and other "rapidly passing fad[s]" that were "in part outright commercial fraud".[1] To a fundamentally nihilistic, retrograde and vain protest he claimed to oppose a vision of the world and a scientific voluntarism that was meant to be superior.

14.2 The Critique of the Mathematic Tendencies of the General Systemology

Another aspect of Bertalanffy's last reflections, in part limited to a private expression in order not to attract enmities, was his ultimate relation with a major trend within the systemic movement which he had initiated.

Even if twenty years earlier he had thought that his general systemology should serve the purpose of a true "logico-mathematical theory", he became in fact in his later years more and more critical with regard to mathematical modelling and axiomatic constructions that invaded systemic research, while they bordered in his view rather on "an indoor sport" than a science, because they did "not refer to any reality", did "not lead to any (actual) discoveries" and "confuse[d] concepts with things". In this regard he criticized, particularly in private, Mihajlo D. Mesarović (1928-) "and the others", who claimed to provide entirely formalized "mathematical theories of general systems", while they did not bring in any knowledge that was not already available about the characteristics of "concrete" systems. As the need for a theoretical point of view did not in any way justify the neglect

---

[1]. (Posthumous) letter from L. von Bertalanffy to P. Cutler (director of the CASBS) (12/06/1972), BCSSS archive.

of empirical problems, such a tendency meant for him a serious danger: namely that systemology ultimately became "a form of modern scholasticism".[2]

14.3 The Disastrous Year of 1971

Bertalanffy described the year of 1971 as "one of the most miserable I have had in my not uneventful life".[3] At the beginning of the year he had a heart attack that forced him to rest three weeks in a hospital. Barely recovered from this ordeal, he returned to SUNY to teach a course entitled "Perspectivist philosophy and the new view of man". The many messages of sympathy that had been sent to the hospital had motivated him to return so rapidly (Davidson, 1983: 67).

Several months later, after a poorly performed cytoscopy, Bertalanffy suffered from a long and terribly painful iatrogenic epididymitis. Reaching the age of seventy in September, he asked the university, as he had done in Edmonton, for an additional year of teaching. This request was acceded to, but simultaneously and just when his pains were the most unbearable, the university informed him that he would not be paid from 1973 onwards. Bertalanffy responded by bringing an action against the university for violating his contract, which stipulated that the university had to pay him as long as he was able to accomplish his teaching, a fitness on which, according to him, nothing cast doubt.[4]

A positive event during that year was an interdisciplinary conference in honour of his 70th birthday,

---

2. Letter from L. von Bertalanffy to A. Locker (11/04/1969). Handwritten note by L. von Bertalanffy that was to be a response to a letter from G. Klir, dated 24/01/1972, BCSSS archive. Translated from the German by ES.
3. (Posthumous) letter from L. von Bertalanffy to G. Tagliacozzo (12/06/72), BCSSS archive.
4. (Posthumous) letter from L. von Bertalanffy to G. Tagliacozzo (12/06/72), BCSSS archive.

which took place on the SUNY campus on September 19. Notably Boulding, Rapoport, Rosen and the philosopher Ervin Laszlo (1932-) were present. The contributions to this conference would be added to those that had already been accumulated for five years in order to form an *oeuvre* of 1,140 pages in his honour and would be published in 1973 (*Unity through Diversity*)—this project had already been formulated on the occasion of his 65th birthday.[5]

14.4 Steps towards a Nomination for the Nobel Prize and Bertalanffy's Last Days

In the spring of 1972 steps were taken in order that Bertalanffy should be nominated for the Nobel prize in physiology. Although he lacked Eccles's support because the latter had already suggested someone else, Bertalanffy won enough support elsewhere so that his file was sent to Oslo.[6] But destiny interfered and crushed this last hope of an acknowledgement of his entire work, finally giving his life a tragic character.

On the morning of June 9, 1972, Bertalanffy worked at home in his office and wrote to the editors of the French, Japanese and Swedish versions of his book *General System Theory*, using his nomination for the Novel prize as a pretext for asking them to accelerate these publications; he did the same with the editor of *Unity through Diversity*.[7]

Still on the morning of June 9, he wrote to the philosopher Jan Kamarýt (1927-), leader of the systemic

---

5. Letter from E. Zerbst to L. von Bertalanffy (01/06/1972), BCSSS archive.
6. B.R. Fuller, quoted in Davidson, 1983: 13.
7. Letters from L. von Bertalanffy to the publishers Dunod, K. Nagano, Strauss, and Gordon & Breach (S. Anderson) (09/06/1972), BCSSS archive.

thought in Czechoslovakia and in charge of the Czech translation of *Robots, Men and Minds*: "I have recently become very interested in the relations of general systems theory and dialectical materialism. Although my knowledge of the latter is limited, it is very interesting to see parallelisms which are even more impressive in view of the completely independent developments"; and he added, "I intend sometime to write about the relations of dialectic materialism and general systems theory, so far as my available energy permits".[8]

These were the last thoughts written by Bertalanffy. Late in the morning of June 9 he suffered a second heart attack and was transported to the intensive care unit. On the evening of June 11 he seemed fresh and alert; he even briefly discussed his various ongoing projects and the people who had to be informed of his condition with his wife. But he died shortly after midnight, on June 12. In accordance with his wishes he was cremated (Davidson, 1983: 69). His ashes rest in the *Côte des Neiges* cemetery in Montreal.

---

8. Letter from L. von Bertalanffy to J. Kamarýt (09/06/1972), BCSSS archive.

Epilogue

Bertalanffy left behind a considerable *oeuvre*. First of all because of its quantity (20 books, including his thesis, and 219 articles), even though it is necessary to put the numbers into perspective, given the recurrence of the same ideas in his publications and the fact that in many of his English articles he revived works he had written in German before 1949.

But it is also and above all considerable because of the diversity of covered problems, ranging from metaphysics and natural philosophy as well as from the critique of culture and ethics, to biology, biophysics, history and philosophy of sciences, philosophy of history, theory of knowledge, history of religions, art history, and the history of stamps and of the postal service. Bertalanffy had, besides Nicholas of Cusa, two major role models of universal thought with whom he largely identified himself: Leibniz and especially Goethe. Even if we let the reader judge the pertinence of this identification and the depth of Bertalanffy's universality by his or her own study of his *oeuvre*, his great erudition, acknowledged even by his critics, is at least undeniable. Judging by his influence in various academic fields and by the remarkable topicality of the problems and ideas he raised (and that, although his name is now largely and, we should add, unjustly forgotten), it is also difficult to deny the fertility of his *oeuvre* and the multiplicity of perspectives it has opened at least in his time.

Finally, Bertalanffy's *oeuvre* is also considerable in that it offers a view of major dialectics in twentieth-century intellectual life that, far from being surpassed in a so-called "post-modernism" or an alleged "end of history", seem hardly to have changed in appearance and, doubtlessly more vivid than ever at the beginning of the twenty-first century, are perhaps all the more dangerous as they are unconscious. The tragedy of Bertalanffy's life and work reflects to a large extent that of his time, but also that of ours. It reveals that the path is shorter than it seemed, the path between the romantic ideal of complete self-fulfilment and the legitimization of the individual's total subordination to social and political logics into which he or she is integrated, between revolutionary pretensions and the dream of a break on the one hand and the most reactionary conservatism on the other, between the feeling of an inevitable decline and the permanent conviction of the advent of a revival, between radical relativism and the quest for the absolute, and between the ideal of a science that liberates the human race and its transformation into an instrument of its enslavement and destruction.

Bertalanffy's life and character can appear extremely appalling. But this does not affect the interest of his thought in any way, whether on a scientific, philosophical or ideological level. And it is only from this point of view that he should be judged. There is little doubt that the quest for a "third way", a constant motif in his works and a reflection of their dialectical dimension, remains topical on all levels of contemporary cultural life.

Finally, it is certainly easier to laugh, in good conscience, at Bertalanffy's reversals of opinion, his tactics, his opportunism or his aristocratism, in particular about his utter contempt of currently dominating values, than to understand their profound origins and reasons and to demonstrate their total lack of justification.

APPENDIX A
The Complete Works of Ludwig von Bertalanffy

(monographs are marked in bold print)

[1923] "Deutsche Mystik," I-III, *Literatur- und Unterhaltungsblatt, Kölnische Zeitung*, 19, 22, 24 December.

[1924] "Einführung in Spenglers Werk," I-IV, *Literatur- und Unterhaltungsblatt, Kölnische Zeitung*, 3, 10, 14, 21 May.

[1924] "Expressionismus und Klassizismus," *Zeitschrift für Ästhetik und allgemeine Kunstwissenschaft*, 18: 334-338.

[1925] "Die Einheit des Bildungstriebes," *Monistische Monatshefte*, 25: 441-445.

[1925] "O. Hagen: Deutsches Sehen—Gestaltungsfragen der deutschen Kunst" (review), *Zeitschrift für Ästhetik und allgemeine Kunstwissenschaft*, 18.

[1926] "Hölderlins Empedokles," *Zeitschrift für Ästhetik und allgemeine Kunstwissenschaft*, 20: 241-248.

[1926] "Die Entdeckung des Raumes," *Zeitschrift für Ästhetik und allgemeine Kunstwissenschaft*, 20: 307-311.

[1926] "Max Dworzak, Kunstgeschichte als Geistesgeschichte" (review), *Zeitschrift für Ästhetik und allgemeine Kunstwissenschaft*, 20: 375-381.

[1926] "Russische Mystik," *Literatur- und Unterhaltungsblatt, Kölnische Zeitung*, 13 April.

[1926] "Zur Theorie der organischen Gestalt," *Roux' Archiv für Entwicklungs-Mechanik*, 108: 413-416.

[1926] **Fechner und das Problem der Integrationen höherer Ordnung**, doctoral thesis, University of Vienna.

[1927] "Das Problem des Lebens," *Scientia*, 41: 265-274.

[1927] "Über die neue Lebensauffassung," *Annalen der Philosophie und philosophischen Kritik*, 6: 250-264.

[1927] "Studien über theoretische Biologie I," *Biologisches Zentralblatt*, 47: 210-242.

[1927] "Über die Bedeutung der Umwälzungen in der Physik für die Biologie. Studien über theoretische Biologie II," *Biologisches Zentralblatt*, 47: 653-662.

[1927] "Eine mnemonische Lebenstheorie als Mittelweg zwischen Mechanismus und Vitalismus," *Biologia Generalis*, 3: 405-410.

[1927] "Die klassische Utopie," *Preußische Jahrbücher*, 210: 341-357.

[1927] "Scientia, Internationale Zeitschrift für wissenschaftliche Synthese" (review), *Preußische Jahrbücher*, 209: 270.

[1928] "Eduard von Hartmann und die moderne Biologie," *Archiv für die Geschichte der Philosophie und Sozialwissenschaft*, 38: 153-170.

[1928] "Philosophie des Organischen (Theoretische Biologie)," *Literarische Berichte aus dem Gebiet der Philosophie*, 17/18: 5-53.

[1928] "J. Strzygowski, Die Krisis der Geisteswissenschaft" (review), *Zeitschrift für Ästhetik und allgemeine Kunstwissenschaft*, 22: 213-220.

[1928] Nikolaus von Kues, Munich, Georg Müller.

[1928] Kritische Theorie der Formbildung, Abhandlungen zur theoretischen Biologie, ed. Julius Schaxel, n° 27, Berlin, Gebrüder Borntraeger (English: *Modern Theories of Development. An Introduction to Theoretical Biology*, Oxford: Clarendon, 1933; New York: Harper Torchbooks 1962).

[1929] "Der heutige Stand des Entwicklungsproblems. Teil I: Die klassischen Theorien; Teil II: Neuere

Anschauungen und die Zukunft der Entwicklungslehre," *Scientia*, 46: 97-110, 171-182.
[1929] "Die Teleologie des Lebens. Eine kritische Erörterung," *Biologia Generalis*, 5: 379-394.
[1929] "Zum Problem der theoretischen Biologie," *Kantstudien*, 34: 374-390.
[1929] "Vorschlag zweier sehr allgemeiner biologischer Gesetze. Studien über theoretische Biologie III," *Biologisches Zentralblatt*, 49: 83-111.
[1929] "Ein Streit um Kant," *Preußische Jahrbücher*, 215: 152-155.
[1929] "Der gegenwärtige Stand des Entwicklungsproblems," *Wissenschaftliche Jahresberichte der Philosophischen Gesellschaft der Universität Wien* (Kant-Society, Vienna), 3-10. (repr. in 102. *Jahresbericht der Schlesischen Gesellschaft für vaterländische Kultur*, 11-16, 1929).
[1929] "Probleme der modernen Metaphysik," *Münchener neuesten Nachrichten*, 28 February or 1 March.
[1929] "Mythos und Wissenschaft. Betrachtungen zur Philosophie des Als-Ob," *Zeitschrift für Menschenkunde*, 4: 329-333.
[1929] "Leben und Energetik," *Unsere Welt*, 21: 214-218.
[1929] "Un Cardinale Germanico (Nicolaus Cusanus)," *Rivista di lettere Scienze ed Arti*, 265: 536-539.
[1929] "E. Becher: Einführung in die Philosophie" (review), *Scientia*, 45: 270.
[1929] "K. Sapper: Philosophie des Organischen" (review), *Scientia*, 45: 273.
[1930] "Neue Richtungen der Entwicklungslehre," *Kosmos—Handweiser für Naturfreunde*, 8: 261-266.
[1930] "Mechanism and vitalism in the light of critical biology. A discussion of the Rignano-Needham controversy," *Psyche Miniatures*, 10: 60-72.

[1930] "Organismische Biologie," *Unsere Welt*, 22: 161-169.

[1930] "R. Wahle - Entstehung des Charakters" (review), *Scientia*, 48: 130-132.

[1930] *Lebenswissenschaft und Bildung*, Erfurt, Kurt Stenger.

[1930/1931] "Tatsachen und Theorien der Formbildung als Weg zum Lebensproblem," *Erkenntnis*, 1: 361-407.

[1931] "Woodger J.H.: Biological principles" (review), *Biologisches Zentralblatt*, 51.

[1931] "Das Vitalismusproblem in ärztlicher Betrachtung," *Medizinische Welt*, 35: 1262-1265.

[1931] "Einleitung," in E. Rignano, *Das Gedächtnis als Grundlage des Lebendigen*, Wien, Braumüller, III-VIII.

[1932] "Vaihingers Lehre von der analogischen Fiktion in ihrer Bedeutung für die Naturphilosophie," in D. Seidel (ed.), *Die Philosophie des Als-Ob und das Leben. Festschrift zu Hans Vaihingers 80. Geburtstag*, Berlin, Reuther & Reichard, pp. 82-91.

[1932] *Theoretische Biologie—I. Band: Allgemeine Theorie, Physikochemie, Aufbau und Entwicklung des Organismus*, Berlin, Gebrüder Borntraeger.

[1933] "Was ist das Leben? Gedanken im Anschluss an neue Forschungsergebnisse über den Bau des Protoplasmas," *Naturforscher*, 10: 117-120.

[1933] "Physikalisch-chemische Theorie des Wachstums," *Biologisches Zentralblatt*, 53: 639-645.

[1933] "Bünning E.: Mechanismus, Vitalismus und Teleologie" (review), *Biologisches Zentralblatt*, 53.

[1933] "Betrachtungen über einige Probleme der Morphologie," *Biologia Generalis*, 9: 70-84.

[1933] "Das organismische Weltbild," *Preußische Jahrbücher*, 234: 252-261.

[1934] "Untersuchungen über die Gesetzlichkeit des Wachstums I. Allgemeine Grundlagen der Theorie.

Mathematisch-physiologische Gesetzlichkeiten des Wachstums bei Wassertieren," *Roux' Archiv für Entwicklungs-Mechanik*, 131: 613-652.

[1934] "Wesen und Geschichte des Homologiebegriffes," *Unsere Welt*, 28: 161-168.

[1934] "Wandlungen des biologischen Denkens," *Neue Jahrbücher für Wissenschaft und Jugendbildung*, 10: 339-366.

[1935] "Gép - ewa Szerverzet," *Búvár: biologicie folyoirat*, 1: 721-723.

[1936] "Az Állatok Fejlödésinék Irányitása," *Búvár: biologicie folyoirat*, 2: 85-89.

[1937] "Biologische Gesetzlichkeit im Lichte der organismischen Auffassung," in *Travaux du IXe Congrès International de Philosophie (Congrès Descartes, 1936)*, Paris, Hermann, 1937, pp. 158-164.

[1937] "Die ganzheitliche Auffassung der Lebenserscheinungen", in M. Sihle and E. Utitz (ed.), *Kongress für synthetische Lebensforschung. Verhandlungsbericht über die Aussprache zwischen Ärzten, Biologen, Psychologen und Philosophen* (Marienbad 1936), Prag, Calve, pp. 100-117.

[1937] Das Gefüge des Lebens, Leipzig, Teubner.

[1938] "A quantitative theory of organic growth. Inquiries on growth laws II," *Human Biology*, 10: 181-213.

[1938] "Studies on the mechanism of growth in Planaria maculata Abstr.," *Biological Bulletin*, 76: 368-369.

[1940] "Der Organismus als physikalisches System betrachtet," *Die Naturwissenschaften*, 28: 521-531.

[1940] "Untersuchungen über die Gesetzlichkeit des Wachstums III. Quantitative Beziehungen zwischen Darmoberfläche und Körpergrösse bei Planaria maculata," *Roux' Archiv für Entwicklungs-Mechanik*, 140: 81-89.

[1940] *Vom Molekül zur Organismenwelt—Grundfragen der modernen Biologie*, Potsdam, Akademische Verlagsgesellschaft Athenaion (2nd edition: 1948).

[1941] "Probleme einer dynamischen Morphologie. Untersuchungen über die Gesetzlichkeit des Wachstums IV," *Biologia Generalis*, 15: 1-22.

[1941] "Wachstumsgradienten und metabolische Gradienten bei Planarien. Untersuchungen über die Gesetzlichkeit des Wachstums V," *Biologia Generalis*, 15: 295-311.

[1941] "Studien zur Reorganisation bei Süsswasserhydrozoen. Untersuchungen über die Gesetzlichkeit des Wachstums VI" (with Rella M.), *Roux' Archiv für Entwicklungs-Mechanik*, 141: 99-110.

[1941] "Stoffwechseltypen und Wachstumstypen. Untersuchungen über die Gesetzlichkeit des Wachstums VII," *Biologisches Zentralblatt*, 61: 511-532.

[1941] "Die organismische Auffassung und ihre Auswirkungen", *Der Biologe*, 10: 247-258, 337-345.

[1942] *Theoretische Biologie—Band II: Stoffwechsel, Wachstum*, Berlin, Gebrüder Borntraeger (2nd edition: Bern, A. Francke AG, 1951).

[1943] "Das Wachstum in seinen physiologischen Grundlagen und seiner Bedeutung für die Entwicklung mit besonderer Berücksichtigung des Menschen," *Zeitschrift für Rassenkunde*, 13: 277-290.

[1943] "Die Abhängigkeit des Stoffwechsels von der Körpergrösse und der Zusammenhang zwischen Stoffwechseltypen und Wachstumstypen. Untersuchungen über die Gesetzlichkeit des Wachstums VIII" (with Müller I.), *Rivista di Biologica*, 35.

[1943] "Der Zusammenhang zwischen Körpergrösse und Stoffwechsel bei Dixippus morosus und

Appendix A: The Complete Works...

seine Beziehung zum Wachstum. Untersuchungen über die Gesetzlichkeit des Wachstums IX" (with Müller I.), *Zeitschrift für vergleichende Physiologie*, 30.

[1943] "Weiteres über die Grössenabhängigkeit des Wachstums. Untersuchungen über die Gesetzlichkeit des Wachstums X" (with Müller I.), *Biologisches Zentralblatt*, 63.

[1943] "Die schädigende Wirkung verschiedener Chinone auf planaria gonocephala DUG. Und ihre Beziehung zur Childschen Gradiententheorie. Untersuchungen über die Gesetzlichkeit des Wachstums XI" (with Schreier O.), *Österreichische zoologische Zeitschrift*.

[1943] "Neue Ergebnisse über Stoffwechseltypen und Wachstumstypen," *Forschungen und Fortschritte*, 19.1/2: 13-15 (English: "Metabolic types and growth types," *Research and Progress*, 9: 279-285, 1943).

[1943] "Jordan P.: Physik und die Geheimnisse des organischen Lebens" (review), *Die Naturwissenschaften*, 31: 23-24.

[1944] "Bemerkungen zum Modell der biologischen Elementareinheiten," *Die Naturwissenschaften*, 32: 26-32.

[1945] "Zu einer allgemeinen Systemlehre," *Blätter für deutsche Philosophie*, 18.3/4 (unpublished).

[1946] "Botanik und Zoologie in Österreich—Aus der Geschichte der Naturwissenschaft," *Universum. Österreichische Monatszeitschrift für Natur, Technik und Wirtschaft*, Gesellschaft für Natur, Technik und Wirtschaft, Vienna, 1: 79-84.

[1946] "A quantitative study of the toxic action of quinines on Planaria gonocephala" (with O. Hoffmann-Schreier and O. Schreier), *Nature*, 158: 948-949.

[1946] "Perspektiven in Kunst und Wissenschaft," *Die Woche. Interessantes aus Wissenschaft und Technik*, 29.

[1946] *Biologie und Medizin*, Vienna, Springer.

[1947] "Vom Sinn und der Einheit der Naturwissenschaften. Aus einem Vortrag von Prof. Dr. Ludwig von Bertalanffy," *Der Student* (Vienna), 2.7/8: 10-11.

[1948] "Das Weltbild der Biologie," in S. Moser (ed.), *Weltbild und Menschenbild, III. Internationale Hochschulwochen des österreichischen College in Alpbach*, Salzburg, Tyrolia, pp. 251-274.

[1948] "Arbeitskreis Biologie," in S. Moser (ed.), *Weltbild und Menschenbild, III. Internationale Hochschulwochen des österreichischen College in Alpbach*, Salzburg, Tyrolia, pp. 355-357.

[1948] "Das organische Wachstum und seine Gesetzmäßigkeiten," *Experientia*, 4: 255-269.

[1948] "Das biologische Weltbild," *Europäische Rundschau*, 17: 782-785.

[1948] "Untersuchungen über bakteriostatische Chinone und andere Antibiotika" (with O. Hoffmann-Ostenhof and O. Schreier), *Monatshefte für Chemie und verwandte Teile anderer Wissenschaften*, 79: 61-71.

[1949] "Zu einer allgemeinen Systemlehre," *Biologia Generalis*, 195: 114-129 (repr. in K. Bleicher (ed.), *Organisation als System*, Wiesbaden, Gabler, 1972, pp. 29-46).

[1949] "Problems of organic growth," *Nature*, 163: 156-158.

[1949] "Goethes Naturauffassung," *Atlantis*, 8: 357-363 (English: "Goethe's concept of nature," *Main Currents in Modern Thought*, 8 (1949): 78-85).

[1949] "Open systems in physics and biology. Ilya Prigogine: Etude thermodynamique des phénomènes irréversibles" (review), *Nature*, 163: 384.

[1949] "Geleitwort des Herausgebers," *Biologia Generalis*, 19: 1-2.

[1949] "Neue Wege zum Lebensproblem," *Atlantis*, 8: 31-32.

[1949] "The concepts of systems in physics and biology," *Bulletin of the British Society for the History of Science*, 1: 44-45.

[1949] "Civilization in the balance," *The Literary Guide and Rationalist Review*, 64: 20-21.

[1949] *Das biologische Weltbild—Die Stellung des Lebens in Natur und Wissenschaft*, Bern, Francke AG. (2nd edition in German: Wien, 1990; English: *Problems of Life: an Evaluation of Modern Biological Thought*, London: Watts & Co, New York: J. Wiley & Sons, 1952. Torch books edition: New York, Harper, 1961. Japanese: Tokyo, Misuzuy Shobo Co., 1954. French: *Les problèmes de la vie*, Paris, Gallimard, 1960. Spanish: *Concepción Biológica del Cosmos*, Faustino, Santiago, Ediciones de la Universidad de Chile, 1963. Dutch: *Het Verschijnsel Leven*, Utrecht, Bijleveld, 1965).

[1950] "An outline of General Systems Theory," *British Journal for the Philosophy of Science*, 1: 139-164.

[1950] "The theory of open systems in physics and biology," *Science*, 111: 23-29 (repr. in F.E. Emery (ed.), *Systems Thinking*, Harmondsworth, Penguin, 1969, pp. 70-85; translated "Teorin om öppna system inom fysik och biologi," in F.E. Emery (ed.), *Systemteori för ekonomer och samhällsvetare*, Stockholm, Stor Prisma, 1972, pp. 17-33).

[1950] "Metabolic types and growth types," *The Anatomical Record. American Society of Zoology*, 108 (123.): 567-568 (repr. in *Revue Canadienne de Biologie*, 10 (1951): 63-64).

[1951] "General System Theory: A new approach to unity of science," 1-6 (with C.G. Hempel, R.E. Bass and H. Jonas), *Human Biology*, 23 (1. "Problems of General Systems Theory", 302-312; 5. "Conclusion", 336-345; 6. "Toward a physical theory of organic teleology—Feedback and dynamics," 346-361).

[1951] "Theoretical models in biology and psychology," *Journal of Personality*, 20: 24-38.

[1951] "Metabolic types and growth types," *American Naturalist*, 85: 111-117.

[1951] "Der Aufstieg der Lebewesen," *Universum, Österreichische Monatszeitschrift für Natur, Technik und Wirtschaft*, 6: 567-571.

[1951] "Tissue respiration and body size" (with W.J.P. Pirozynski), *Science*, 113: 599-600 and 114: 306-307.

[1951] "Comments and communications. Tissue respiration and body size," *Science*, 114: 307.

[1951] *Auf den Pfaden des Lebens—Ein biologisches Skizzenbuch*, Frankfurt/Main, Umschau Verlag.

[1952] "On the logical status of the theory of evolution," *Laval Théologique et Philosophique*, 8: 161-168.

[1952] "Ontogenetic and evolutionary allometry" (with W.J.P. Pirozynski), *Evolution*, 6: 387-392.

[1952] "Planarians as model organisms for morphogenesis and pharmaco-dynamical actions," *Revue Canadienne de Biologie*, 11: 54.

[1952] "Is the rate of basal metabolism determined by tissue respiration?" (with W.J.P. Pirozynski), *Revue Canadienne de Biologie*, 11: 77-78.

[1952] "Ribonucleid acid in cytoplasm of liver cells: Its localization in Hyperplasia and Hepatoma produced by 2-acetylaminofluorene" (with W.J.P. Pirozynski), *Archives of Pathology*, 54: 450-457.

[1953] "Philosophy of science in scientific education," *Scientific Monthly*, 77: 233-239.

[1953] "Tissue respiration, growth and basal metabolism" (with W.J.P. Pirozynski), *Biological Bulletin*, 105: 240-256.

[1953] "Tissue respiration of musculature in relation to body size" (with R.R. Estwick), *American Journal of Physiology*, 173: 58-60.

[1953] "The surface rule in crustaceans" (with J. Krywienczyk), American *Naturalist*, 87: 107-110.

[1953] "Effects of Hormones on the Distribution of Ribonucleid Acid in Liver Cells: Changes Following Administation of Cortisone, Desoxycorticosterene Acetate and Thyroxine" (with W.J.P. Pirozynski), *Acta Anatomica*, 19: 7-14.

[1953] ***Biophysik des Fließgleichgewicht—Einführung in die Physik offener Systeme und ihre Anwendung in der Biologie***, Braunschweig, Vieweg & Sohn (2nd rev. ed. with Beier W. & Laue R., 1977).

[1954] "The biophysics of the steady state of the organism," *Scientia*, 48: 361-365.

[1954] "Biophysik auf neuen Bahnen," *Naturwissenschaftliche Rundschau*, 7: 418-420.

[1954] "Rudolf Virchow, 1821-1902," *The Canadian Medical Association Journal*, 70: 581.

[1954] "Das Fließgleichgewicht des Organismus," *Kolloid-Zeitschrift*, 139: 86-91.

[1954] "A discussion of the psychophysical problem" (with K.W. Deutsch), in R. Grinker (ed.), *Proceedings of the 7th Conference on the Unified Theory of Human Nature*, Chicago, Michael Reese Hospital (mimeograph).

[1954] "Tissue respiration in experimental and congenital pituitary deficiency" (with R.R. Estwick), *American Journal of Physiology*, 117: 16-18.

[1955] "General Systems Theory," *Main Currents in Modern Thought*, 11: 75-83 (repr. in *General Systems*, I (1956): 1-10; in R.W. Taylor (ed.), *Life, Language, Law. Essays in Honor of A. F. Bentley*, Yellow Springs, Antioch Press, 1957, pp. 58-78; in J.D. Singer (ed.), *Human Behavior and International Politics: Contributions from the Social-Psychological Sciences*, Chicago, Rand McNally & Co., 1965, pp. 10-31).

[1955] "An essay on the relativity of categories," *Philosophy of Science*, 225: 243-263 (repr. in *General Systems*, 7 (1962): 71-83).

[1955] "Die Evolution der Organismen," in J. Schlemmer, *Schöpfungsglaube und Evolutionstheorie*, Stuttgart, Kröner, pp. 53-66.

[1955] "Correlation of O_2—Consumption with body size: invertebrates," in E.C. Albritten (ed.), *Standard Values in Nutrition and Metabolism*, Tables 137, 230, 357 (rev. table with Locker A., in P.L. Altmann and D.S. Dittmer (ed.), *Metabolism, Biological Handbooks*, FASEB Federation of American Societies for Experimental Biology, Bethesda, Maryland, 1968, Tables 67, 372-377).

[1955] "Changes of cytoplasmic basophilia during carcinogenesis induced by 2-acetylaminofluorene" (with W.J.P. Pirozynski), *Experimental Medicine and Surgery*, 13: 261-269.

[1956] "A biologist looks at human nature," *Scientific Monthly*, 82: 33-41 (repr. in D.S. Robert (ed.), *Contemporary Readings in General Psychology*, Boston, Houghton Mifflin, 1959 (1st ed.) and 1965 (2nd ed.), pp. 267-275; in S.J. Beck and H.B. Molish (ed.), *Reflexes to Intelligence. A Reader in Clinical Psychology*, New York-Glencoe, The Free Press, 1959, pp. 629-640).

[1956] "The principle of allometry in biology and the social sciences" (with R.S. Naroll), *General Systems*, 1: 76-89.

[1956] "Some considerations on Psychobiological Development", Paper read at the Study Groups on the Psychobiological Development of the Child, *World Health Organisation*, Geneva, WHO/AHP/11.

[1956] "Die Beiträge der Systemtheorie zum Wissenschaftsdenken der Gegenwart", *RIAS, Funk-Universität,* Radio broadcast on Tuesday, 2 October, Berlin.

Appendix A: The Complete Works...

[1956] "Some considerations on growth in its physical and mental aspects," *Merrill-Palmer Quarterly*, 3: 13-23.

[1956] "Das Monopol des Menschen—Vom Sinn unserer biologischen Existenz," *Stuttgarter Zeitung* (Sunday supplement), 1.12.

[1956] "Moderne Hypothesen für die Entstehung des Lebens," *Kosmos*, 52: 255-260 (repr. in O.W. Haselhoff and H. Stachowiak (ed.), *Schriften zur wissenschaftlichen Weltorientierung, Vol. V: Stammesgeschichte, Umwelt, Menschenbild*, Berlin, Lüttke, 1959, pp. 7-18).

[1956] "Identification of cytoplasmic basophilia (Ribonucleid Acid) by fluorescence microscopy" (with I. Bickis), *Journal of Histochemistry and Cytochemistry*, 4: 481-493.

[1956] "Use of acridine-orange fluorescence technique in exfoliative cytology" (with F. Masin and M. Masin), *Science*, 124: 1024-1025.

[1957] "Moderne Forschung und Wissenschaftsbetrieb," *Deutsche Universitätszeitung*, 12: 4-5.

[1957] "Allgemeine Systemtheorie. Wege zu einer Mathesis universalis," *Deutsche Universitätszeitung*, 12: 8-12.

[1957] "Mutation und Evolution," in J. Schlemmer (ed.), *Genetik - Wissenschaft der Entscheidung. Das Heidelberger Studio. Eine Sendereihe des Süddeutschen Rundfunks*, Stuttgart, Kröner, pp. 103-116.

[1957] "The significance of psychotropic drugs for a theory of psychosis," Paper read at the study group on ataractics and hallucinogenics, *World Health Organization*, Geneva, WHO/AHP/2, pp. 1-37 (published in *Ataractic and hallucinogenic drugs in psychiatry. Report of a Study Group*, Geneva, WHO-Technical Report Series, No. 152, 1957).

[1957] "Psychobiological development of the child," *Science*, 125: 125.

[1957] "Semantics and General System Theory," *General Semantics Bulletin*, 20/21: 41-45.

[1957] "Wachstum," in *Kükenthals Handbuch der Zoologie*, vol. 8, 4 (6), Berlin, De Gruyter, pp. 1-68 (rev. English translation: "Principles and theory of growth," in W.W. Nowinski (ed.), *Fundamental Aspects of Normal and Malignant Growth*, Amsterdam, Elsevier, 1960, pp. 137-259).

[1957] "La Teoria generale dei Sistemi," *La Voce dell'America*, 18-G u. 2-H.

[1957] "Neue Wege der Biophysik," *Physikalische Verhandlungen*, 8: 5-6.

[1957] "Quantitative laws in metabolism and growth," *Quarterly Review of Biology*, 32: 217-231.

[1957] "Detection of gynecological cancer: Use of fluorescence microscopy to show nucleid acids in malignant growth" (with F. Masin, M. Masin and L. Kaplan), *California Medicine*, 87: 248-251.

[1958] "Die biologische Sonderstellung des Menschen," in J. Schlemmer (ed.), *Die Freiheit der Persönlichkeit*, Das Heidelberger Studio. Eine Sendereihe des Süddeutschen Rundfunks, Stuttgart, Kröner, pp. 7-21.

[1958] "Comments on aggression," *Bulletin of the Menninger Clinic*, 22: 50-57 (repr. in I.G. Sarason (ed.), *Psychoanalysis and the Study of Behavior*, Princeton, NJ, Nostrand, 1965, pp. 107-116).

[1958] "Human values in a changing world," in *Science and Religion as Approaches to Reality*, Second Annual Conference on Science and Religion, Palo Alto, pp. 1-8 (repr. in A.H. Maslow (ed.), *New Knowledge in Human Values*, New York, Harper & Brothers, 1959, pp. 65-74).

[1958] "Fluorescence microscopy in cancer diagnosis (Abstract)" (with F. Masin, M. Masin and L. Kaplan), *American Academy of General Practice*, Abstracts of the annual scientific assembly (Kansas City), pp. 208-211.

[1958] "A new and rapid method for diagnosis of vaginal and cervical cancer by fluorescence microscopy" (with F. Masin and M. Masin), *Cancer*, 11: 873-887.

[1959] "Modern concepts on Biological Adaptation," in C. Mc Chandler Brooks and P.F. Cranefield (ed.), *The Historical Development of Physiological Thought*, New York, Hofner, pp. 265-286.

[1959] "Eine fluoreszenzmikroskopische Schnellmethode zur Diagnose des gynäkologischen Carcinoms," *Klinische Wochenschrift*, 37: 469-471.

[1959] "Cancer diagnosis by fluorescence microscopy," *Modern Medicine*, 112-113.

[1959] "Fluorescence microscopy of irradiated cells," *Acta Cytologia*, 3: 354, 361, 367 (repr. in *Acta Cytologica*, 8, 1964).

[1959] "Some biological considerations on the problem of mental illness," *Bulletin of the Menninger Clinic*, 23: 41-55 (repr. in L. Appleby, J. Scher and J. Cummings (ed.), *Chronic Schizophrenia. Exploration in theory and treatment*, Glencoe, The Free Press, 1960, pp. 36-53).

[1959] "A fluorescence-microscopic method of cancer detection in bronchogenic cancers" (with F. von Bertalanffy), *Proceedings of the 2nd Workshop Conference on Lung Cancer Research*, American Cancer Society, Philadelphia (Proceedings of the National Cancer Conference), pp. 97-100.

[1959] "Cytological cancer diagnosis. A new approach based upon acridine orange fluorescence microscopy" (with F. von Bertalanffy), *What's new (Abbott Laboratories)*, 214: 7-14 (Canadian ed. no. 125 (Spring 1960): 9-12).

[1960] "The psychopathology of scientism," in H. Schoeck and J.W. Wiggins (ed.), *Scientism and Values*, Princeton, NJ, Nostrand, pp. 202-218.

[1960] "Allgemeine Systemtheorie und die Einheit der Wissenschaften," Atti del XII Congresso Internazionale di Filozofia, 5, Firenze, pp. 55-61.

[1960] "Neue Wege biologisch-medizinischen Denkens," Festvortrag, Ärztliches Collegium, Regensburg (repr. in *Ärztliche Mitteilungen*, 46 (1961): 2389-2396).

[1960] "A new method for cytological diagnosis of pulmonary cancer" (with F. von Bertalanffy), *Annals of the New York Academy of Sciences*, 84: 225-238.

[1960] "General System Theory and the behavorial sciences," in J.M. Tanner and B. Inhelder (ed.), *Discussions on Child Development*, 4th Meeting of the World Health Organization. Study Groups on the Psychobiological Development of the Child, vol. 4, London, Tavistock, pp. 155-175.

[1960] "Fluorescence microscopy in the study of nucleo-cytoplasmic relations," *Symposium on Germ Cells and Development*, Institute Intern d'Embryologie and Foundazione A Baselli (Milano), p. 145.

[1960] "Die Fluoreszenzmethode in der exfoliativen Cytologie, besonders die Diagnose des Lungenkrebses" (with F. von Bertalanffy), *Die Naturwissenschaften*, 47: 165-166.

[1960] "Fluorescence microscopy of cervical cells and macrophages" (with F. von Bertalanffy), *Acta Cytologia*, 4: 298.

[1960] "Acridine orange fluorochrome in the study of normal and malignant epithelium of the uterine cervix" (with F. Masin, M. Masin, L. Kaplan and R. Carleton), *American Journal of Obstetrics and Gynecology*, 80: 1063-1073.

[1961] "Fluorescence microscopy in the study of nucleo-cytoplasmic relations," *Symposium on Germ Cells and Development*, Institute Intern. d'Embryologie and Fondazione A Baselli, pp. 145-146.

[1961] "Die Fluoreszenzmethode in der Zytodiagnostik des gynäkologischen Karzinoms" (with F. von Bertalanffy), *Die Medizinische Welt*, 35: 1742-1751.

Appendix A: The Complete Works...

[1961] "Fluorescence microscopy of hyperplasia in gynecological cytodiagnosis" (with F. von Bertalanffy and A.M. Goodwin), *Acta Cytologia*, 5: 256-257.

[1961] "Normal processes in cervical and vaginal epithelia and their implications in malignant growth" (with F. von Bertalanffy), *Acta Cytologia*, 5: 302-305.

[1962] "General System Theory - A critical review," *General Systems*, 7: 1-20 (repr. in W. Buckley (ed.), *Modern Systems Research for the Behavorial Scientist*, Chicago, Aldine Publishing Co., 1968, pp. 11-30; in M.M. Milstein and J.A. Belasco (ed.), *Educational Administration and the Behavioral Sciences*, Boston, Allyn & Bacon, 1973, pp. ix-xi and 5-49).

[1962] "Modern paths of biologico-medical thought," *Yale Scientific Magazine*, Dec.

[1962] "Democracy and elite: The educational quest," *Main Currents in Modern Thought*, 195: 31-36.

[1962] "Fluorescence cytodiagnosis—A way toward expansion of early cancer detection," *The Alberta Medical Bulletin*, 27: 94-98.

[1962] "Cytodiagnosis of cancer by fluorescence microscopy," *Annales d'Histochimie*, 7, suppl. 1, 85-94.

[1962] "Fluorescence spots cancer cells, RNA for easy diagnosis," *Medical Tribune*, 3: 20.

[1962] "Recent advances in fluorescence cytodiagnosis of cancer," *International Journal of Cancer*, 1329-1330 (complete translation: "Akridinorange-Fluoreszenz-Cytodiagnostik in der Früherkennung des Krebses" (with F. von Bertalanffy), *Ärztliche Mitteilungen*, 47 (1962): 2393-2397).

[1962] "Leeser, O.: Neue Wege biologisch-medizinischen Denkens" (review), *Ärztliche Mitteilungen*, 47: 2040-2046.

[1962] "The origin of posts. Italy, early 15th century," *The Philatelist and Postal Historian*, Suppl., New York, Dec.

[1963] "Acridine orange fluorescence in cell physiology, cytochemistry and medicine," *Protoplasma*, 57: 51-83.

[1963] "Venetia - 1390-1797. Commerce and sea mail of the Venetian Republic," *Postal History Journal*, 7: 17-32, (Italian translation: "Venezia 1390-1797. Il Commercio e la Posta di Mare della Republica di Venezia," *Filatelia* (Roma), 69 (1969): 34-43).

[1963] "Sardinia: Decrees introducing the first postal stationery (1818-1819)," *Postal History Journal*, 7.

[1964] "Basic concepts in quantitative biology of metabolism", in *Quantitative Biology of Metabolism-First International Symposium*, Helgoländer wissenschaftliche Meeresuntersuchungen, 9: 5-37.

[1964] "The mind-body problem: A new view," *Psychosomatic Medicine*, 26: 29-45 (repr. in F.W. Matson and A. Montagu (ed.), *The Human Dialogue. Perspective on Communication*, New York, Free Press, 1967, pp. 224-245).

[1964] "Biophysics of open systems," *Gemeinsame Tagung der Deutschen Gesellschaft für Biophysik e.V., der Österreichischen Gesellschaft für reine und angewandte Biophysik und der Schweizerischen Gesellschaft für Strahlenbiologie*, Wien, 14.-16.9.1964, Tagungsbericht: Comp. Weiner, 1-9 (German translation: "Die Biophysik offener Systeme," *Naturwissenschaftliche Rundschau*, 18 (1965): 467-469).

[1964] "The World of science and the world of value," *Teachers College Record*, 65: 244-255 (repr. in J.T. Bugental (ed.), *Challenges of Humanistic Psychology*, New York, McGraw Hill, 1967, pp. 335-344; in E. Shoben and S. Goldberg (ed.), *Problems and Issues in Contemporary Education*, Glenview, Scott, 1968; in E.H. Odell (ed.), *A College Looks at American Values*, Ellensburg, Central Washington State College, 1971, pp. 9-21).

Appendix A: The Complete Works...

[1964] "Der gegenwärtige Stand der Fluoreszenz-Zytodiagnose des Karzinoms," in W. Bickenbach and H.J. Soost (ed.), *Berichte über die I. Tagung der Deutschen Gesellschaft für Angewandte Zytologie*, München, Müller, pp. 177-183.

[1964] "The consultant," *Acta Cytologia*, 8: 377-380.

[1964] "Recent advances of fluorescence cytodiagnosis of cancer," *Acta Unio Internationalis Contra Cancrum*, International Union against Cancer, Louvain, 20: 1329-1330.

[1964] "Cognitive processes and psychopathology," Address at *Symposium of the Academy of Psychoanalysis*, Montreal.

[1965] "Zur Geschichte theoretischer Modelle in der Biologie", *Studium Generale*, 18: 290-298.

[1965] "On the definition of symbol," in J.R. Royce (ed.), *Psychology and the Symbol. An Interdisciplinary Symposium*, New York, Random House, pp. 26-72.

[1965] "General System Theory and psychiatry," in S. Arieti (ed.), *American Handbook of Psychiatry*, vol. 3, New York, Basic Books, pp. 705-721 (2nd ed., vol. 1: pp. 1095-1117, 1974).

[1965] "Professor Bernhard Rensch zum 65. Geburtstag," Beilage zur *Naturwissenschaftliche Rundschau, 18. Mitteilungen des Verbandes Deutscher Biologen*, Heft 1, pp. 482-484.

[1966] "Biologie und Erforschung des Lebens," *Bild der Wissenschaft*, 3: 708-719.

[1966] "Mind and body re-examined," *Journal of Humanistic Psychology*, 6: 113-138.

[1966] "Histoire et méthodes de la théorie générale des systèmes", *Atomes*, 21 : 100-104.

[1966] "On the von Bertalanffy growth curve. Rectification of an error, and suggestions for further use of the equation under consideration," *Growth*, 30: 123-124.

[1966] "The tree of knowledge," in G. Kepes (ed.), *Sign, Image, Symbol*, New York, Braziller, pp. 274-278.

[1967] "General Theory of Systems: Application to psychology," *Social Science Information*, 6: 125-136 (repr. in *The Social Sciences: Problems and Orientation*, Paris, La Haye, Mouton UNESCO, 1968, pp. 309-319).

[1967] "General Systems Theory and Psychiatry: An Overview," *American Psychiatric Association* 176, Annual Meeting (published in *Psychiatric Spectator*, 4, 6-8, 1967; in W. Gray, N.D. Rizzo and F.J. Dahl (ed.), *General Systems Theory and Psychiatry*, Boston, Little, Brown & Company, 1969, pp. 33-46).

[1967] "Origine delle Poste. L'Italia Agli Inizi Del Secolo XV," *Filatelia*, 41.25: 26-32.

[1967] Robots, Men and Minds, New York, Braziller (German: ... *aber vom Menschen wissen wir nichts*, Düsseldorf, Wien, Econ Verlag, 1970; French: *Des Robots, des esprits et des hommes*, Paris, Dunod, 1972; Italian: *Il Sistema Uomo—La psichologia nel mondo moderno*, Milano, Etas-Kompass, 1971; Spanish: *Robots, Hombres y Mentes*, Madrid, Guadarrama: 1971; Czech: Praha, Svoboda, 1972; Japanese: Tokyo, Misuzu Shobo, 1971).

[1968] "Symbolismus und Anthropogenese," in B. Rensch (ed.), *Handgebrauch und Verständigung bei Affen und Frühmenschen, Symposium der Werner-Reimers-Stiftung für anthropogenetische Forschung*, Bern-Stuttgart, pp. 131-148.

[1968] "Das Modell des offenen Systems," *Nova Acta Leopoldina*, 33: 73-87.

[1968] "General Systems Theory and a new view of the nature of man" (review), American Psychiatric Association, Annual Meeting, *Psychiatric Spectator*, 5: 13-14.

[1968] "General Systems Theory as integrating factor in contemporary science and in philosophy," *Akten des XV. Internationalen Kongresses für Philosophie*, 2, Wien 1969, pp. 335-340 (repr. in *FORUM*, Publication of the International Center for Integrative Studies, New York, 1, 1969, pp. 33-38).

[1968] *General System Theory—Foundations, Development, Applications*, New York, Braziller (New York, Edition of Professional and Technical Programs, Inc., 1968; New York, Book Find Club, 1969; German: Braunschweig, Vieweg, 1971; British: London, Penguin, 1971; French: Paris, Dunod, 1973; Italian: Milano, Etas Kompass, 1972; Spanish: Madrid, Guadarrama, 1972; Swedish: Stockholm, Wahlshom & Widstrand, 1972; Japanese: Tokyo, Misuzu Shobo, 1972).

[1968] *Organismic Psychology and System Theory*, Worcester, Clark University Press (Italian: Roma, Armando, 1973).

[1968] *Kurzlehrbuch für Biologie* (with P. Lüth), München, Lehmann.

[1968] *The Living Flame. Collected Essays 1924-1967*, New York, Braziller.

[1968] *History of the Letter. From Early Times to 1840*, New York, Braziller.

[1968] *General Systems Theory. A Reader* (with **R.W. Jones and A. Rapoport**), New York, Academic Press.

[1969] "Evolution: Chance or Law," in A. Koestler and J.R. Smithies (eds.), *Beyond Reductionism. The Alpbach Symposium. New Perspectives in the Life Sciences*, London, Hutchinson, pp. 56-84.

[1969] "Gefügegesetzlichkeit," in J. Josef Ritter (ed.), *Historisches Wörterbuch der Philosophie*, vol. I, Basel, pp. 80-82.

[1969] "Venezia 1390-1797. Il commercio e la Posta di Mare della Republica di Venezia," *Filatelia*, 69.25: 34-43.

[1970] "General System Theory and psychology," in J.R. Royce (ed.), *Toward Unification of Psychology*, Toronto, University Press, pp. 220-223.

[1970] "Leben," in D. Kernig (ed.), *Sowjetsystem und Demokratische Gesellschaft. Eine vergleichende Enzyklopädie*, Freiburg-Basel-Wien, Herder, pp. 1373-1384.

[1970] "Biologie und Weltbild," in M. Lohman (ed.), *Wohin führt die Biologie? Ein interdisziplinäres Kolloquium*, München, Carl Hanser Verlag, pp. 13-31.

[1971] "Cultures as systems—Toward a critique of historical reason" (Paper presented at the Annual Meeting of the American Historical Association, New York, 29.12.1971), *Bucknell Review*, 22: 151-161 (repr. in R. Garvin (ed.), *Phenomenology, Structuralism, Semiology*, Cranbury, Bucknell Review, Associated University Press, 1976, pp. 151-161).

[1971] "System, symbol and the image of man (Man's immediate socio-ecological world)," in Galdston (ed.), *The Interface Between Psychiatry and Anthropology*, New York, Brunner-Mazel, pp. 88-119.

[1971] "Body, mind, and values," in E. Laszlo and J.B. Wilbut (ed.), *Human Values and the Mind of Man. Proceedings of the Fourth Conference on Value Inquiry*, London-New York, Gordon and Breach, pp. 33-47.

[1971] "Progress in General System Theory," in *Proceedings of the XIIIth International Congress of the History of Science*, Section 1A, Moscow, pp. 18-22.

[1971] "Science and the world of value", in E.H. Odell (ed.), *A College Looks at American Values*, Ellensburg, Central Washington State College.

[1972] "The History and Status of General Systems Theory," in G. Klir (ed.), *Trends in General Systems Theory*, New York, Wiley, pp. 21-41 (repr. in J.D. Couger and R. W. Knapp (ed.), *System Analysis*

Techniques, New York, Wiley, 1974, pp. 9-26; in *Academy of Management Journal*, 15.4 (1972): 407-426; in R.L. Ackoff (ed.), *Systems and Management Annual 1974*, New York, Petrocelli Books, 1974, pp. 3-23; Russian translation: "Istorija i status obˇs ˇcej teorii sistem, Sistemnye issledovanija," 1973, 23-38).

[1972] "Vorläufer und Begründer der Systemtheorie," in R. Kurzrock (ed.), *Systemtheorie. Forschung und Information*, Schriftenreihe der RIAS-Funkuniversität, Berlin, Colloquium Verlag, pp. 17-27.

[1972] "Humanism and Antihumanism in the present age," *The Humanist*, Sept./Oct., 14.

[1972] "The quest for systems philosophy," *Metaphilosophy*, 3.2: 142-145.

[1972] "Response," in E. Laszlo (ed.), *The Relevance of General Systems Theory—Papers presented to L. von Bertalanffy on his 70th birthday*", New York, Braziller, pp. 181-191.

[1972] "The model of open systems: Beyond molecular biology," in A.D. Breck and W. Yourgrau (ed.), *Biology, History and Natural Philosophy*, New York, Plenum, pp. 17-30.

[1972] "Symposium on robots, men, and minds. Ludwig von Bertalanffy," *The Philosophy Forum*, 9: 301-329.

[1972] "A biologist looks at human nature—Reconsideration 1972: A mini-history of the concept of symbolism," *Quarterly Bulletin, Center for Theoretical Biology*, State University of Buffalo at New York, pp. 153-161.

[1972] "Foreword", in E. Laszlo (ed.), *Introduction to Systems Philosophy*, New York, Gordon & Breach, pp. xvii-xxi.

Posthumous Publications

[1973] "The history of the letter from the late Middle Ages. An Introduction to five centuries of communication," *Postal History Journal*, 17: 1-41.

[1974] "The unified theory for psychiatry and the behavioral sciences," in S.C. Feinstein and P.L. Giovacchini (ed.), *Adolescent Psychiatry*, New York, basic Books, vol. 3, pp. 43-48.

[1974] "Gefüge und Homöostase," in J. Ritter (ed.), *Historisches Wörterbuch der Philosophie*, Basel and Stuttgart, Schwabe & Co., vol. 3, pp. 80-82, 1184-1186.

[1975] *Perspectives on General System Theory— Scientific-Philosophical Studies,* Edgar Taschdjian (ed.), New York, Braziller.

[1977] "The role of systems theory in present day science, technology and philosophy," in K.E. Schaefer, H. Hensel and R. Brady (ed.), *Toward a Man-Centered Medical Science*, Mt. Kisco, New York, Futura, pp. 11-15.

Appendix A: The Complete Works...

Chief Editor of:

Handbuch der Biologie **(with F. Gessner)**, 14 volumes, 80 co-authors, Konstanz, Akademische Verlagsgesellschaft Athenaion, (1942-1967).
General Systems **(with A. Rapoport)**, 28 volumes, Washington, Society for General Systems Research (1956-1972); the collection was continued until 1983/1984.

Co-Editor of:

Biologia Generalis
Fortschritte der experimentellen und theoretischen Biophysik
Main Currents in Modern Thought
The Philosophy Forum
Études d'Épistémologie génétique

APPENDIX B
Secondary Sources on Ludwig von Bertalanffy's Life and Thought

(Excluding works not primarily or largely based on a discussion of Bertalanffy's ideas. Underlining indicates works that refer to details of Bertalanffy's life.)

Afanasjew, W.G. (1962). "Über Bertalanffys *organismische Konzeption*," *Deutsche Zeitschrift für Philosophie*, 8.

Ashby, W.R. (1955-1956). "Biophysik des Fliessgleichgewicht" (review), *British Journal of Philosophy of Science*, 6.

Ballauff, T. (1940). "Über das Problem der autonomen Entwicklung im organischen Seinsbereich," *Blätter für deutsche Philosophie*, 14.

Ballauff, T. (1943). "Die gegenwärtige Lage der Problematik des organischen Seins," *Blätter für deutsche Philosophie*, 17.

Bass, R.E. (1951). "Unity of nature," *Human Biology*, 23.

Bavink, B. (1930). "Besprechungen über Bertalanffy," *Unsere Welt*, (January).

Bavink. B. (1933). "Review of Theoretische Biologie," *Unsere Welt*, 25.

Bello, R. (1985). "The Systems Approach—A. Bogdanov and L. von Bertalanffy," *Studies in Soviet Thought*, 30.2.

Bendmann, A. (1963). "Die organismische Auffassung Bertalanffys," *Deutsche Zeitschrift für Philosophie*, 11.

Bendmann, A. (1967). *L. von Bertalanffys organismische Auffassung des Lebens in ihren philosophischen Konsequenzen*, Jena, Gischer.

Bendmann, A. (1973). "Materialism and biology today," in W. Gray and N.D. Rizzo (ed.), *Unity through Diversity—A Festschrift for Ludwig von Bertalanffy*, vol. 1, New York, Gordon & Breach.

Berlinski, D. (1978). "Adverse notes on systems theory," in J. Klir (ed.), *Applied General Systems Research Recent Development and Trends*, New York: Plenum Press, 1978, vol. 5, pp. 949-960.

Bertalanffy, Maria von (1973). "Reminiscences", in W. Gray and N.D. Rizzo (ed.), *Unity through Diversity—A Festschrift for Ludwig von Bertalanffy*, vol. 1, New York, Gordon & Breach, pp. 31-52.

Brauckmann, S. (1996). "The organismic system theory of L. von Bertalanffy," *Biologisches Zentralblatt*, 115.

Brauckmann, S. (1997). *Eine Theorie für Lebendes? Die Synthetische Antwort Ludwig von Bertalanffys*, doctoral thesis, University of Münster, published Frankfurt am Main, Munich, New York, Hänsel-Hohenhausen, 2000, ISBN 3-8267-2682-0.

Brauckmann, S. (2000). "The organism and the open system—Ervin Bauer and Ludwig von Bertalanffy," *Annals of the New York Academy of Sciences*, 901: 291-300.

Bünning, E. (1949). "Das biologische Weltbild" (review), *Die Naturwissenschaften*, 36: 287-288.

Clark, J.W. (1972). "The general ecology of knowledge in curriculums of the future," in E. Laszlo (ed.), *The relevance of general systems theory*, New York, Braziller.

Chroust, G. and Hofkirchner, W. (2006). "Ludwig von Bertalanffy returns home", in *Systems Research and Behavioral Science*, 23: 1-3.

Davidson, M. (1983). *Uncommon Sense: the Life and Thought of Ludwig von Bertalanffy, Father of the General Systems Theory*, Los Angeles, J.P. Teacher.

Dehlinger, U. and Wertz, E. (1942). "Biologische Grundfragen in physikalischer Betrachtung," *Naturwissenschaften*, 30: 250-253.

De Vajay, S. (1973). "The Bertalanffys. Their lineage within the social structure of Hungary," in W. Gray and N.D. Rizzo (ed.), *Unity through Diversity—A Festschrift for Ludwig von Bertalanffy*, vol. 1, New York, Gordon & Breach., pp. 12-20.

Donath, T. (1973). "Von Bertalanffy's integrative endeavors," in W. Gray and N.D. Rizzo (ed.), *Unity through Diversity—A Festschrift for Ludwig von Bertalanffy*, vol. 1, New York, Gordon & Breach.

Dotterweich, H. (1940). *Das biologische Gleichgewicht und seine Bedeutung für die Hauptprobleme der Biologie*, Jena.

Drack, M., Apfalter W. and Pouvreau D. (2007). "On the making of a system theory of life: Paul A. Weiss and Ludwig von Bertalanffy's conceptual connection," *Quarterly Review of Biology*, 82.4: 349-374.

Dubrovsky, V. (2004). "Toward system principles: general system theory and the alternative approach," *Systems Research and Behavioral Science*, 21.2: 109-122.

Egler, F.E. (1953). "Bertalanffian organismicism," *Ecology*, 34.

Eisikovits, R.A. (1984). "Descartes and Bertalanffy—Break or continuity?," *Journal of Thought*, 19.1.

Eugene, J. (1981). *Aspects de la théorie générale des systèmes: une recherche des universaux*, Paris, Maloine.

Fries, C. (1935). "Wiedergeburt der Naturphilosophie," *Geistige Arbeit*, 7.

Fries, C. (1936). *Metaphysik als Naturwissenschaft. Betrachtungen zu Ludwig von Bertalanffys Theoretischer Biologie*, Berlin.

Georgiou, I. (1999). "Groundwork of a Satrean input toward informing some concerns of critical systems thinking," *Systemic Practice and Action Research*, 12.6: 585-605.

Georgiou, I. (2000). "The ontological constitution of bounding-judging in the phenomenological epistemology of Von Bertalanffy's general system theory," *Systemic Practice and Action Research*, 13.3: 391-424.

Gessner, F. (1934). "Theoretische Biologie" (review), *Freie Welt* (Gablonz), 14.

Gray, W.D. (1969). "Review of LvB's *Robots, men and minds*," *General Systems*, 14.

Gray, W. (1971). "Ludwig von Bertalanffy's general system theory as a model for humanistic system science," in *Proceedings of the 13th International Congress of the History of Science*, Section 1A, Moscow.

Gray, W. (1972). "Bertalanffian principles as a basis for humanistic psychiatry," in E. Laszlo (ed.), *The relevance of general systems theory*, New York, Braziller.

Gray, W. (1973). "Introduction," in W. Gray and N.D. Rizzo (ed.), *Unity through Diversity—A Festschrift for Ludwig von Bertalanffy*, vol. 2, New York, Gordon & Breach.

Gray, W. (1973) "LvB and the development of modern psychiatric thought," in W. Gray and N.D. Rizzo (ed.), *Unity through Diversity—A Festschrift for Ludwig von Bertalanffy*, vol. 1, New York, Gordon & Breach.

Gray, W. and Rizzo, N.D. (ed.) (1973). *Unity through Diversity—A Festschrift for Ludwig von Bertalanffy*, New York, Gordon & Breach.

Gray, W. and Rizzo, N.D. (1973). "In memoriam & Introduction," in W. Gray and N.D. Rizzo (ed.), *Unity through Diversity—A Festschrift for Ludwig von Bertalanffy*, vol. 1, New York, Gordon & Breach.

Grinker, R.R. Sr. (1974). "In memory of LvB's contribution to psychiatry," *General Systems*, 19.

Groß, J. (1930). "Die Krisis in der theoretischen Physik und ihre Bedeutung für die Biologie," *Biologisches Zentralblatt*, 50.

Appendix B: Secondary Sources...

Hammond, D. (2003). *The Science of Synthesis—Exploring the Social Implications of General Systems Theory*, University Press of Colorado.

Hempel, C.G. (1951). "General systems theory and the unity of science," *Human Biology*, 23.

Hofer, V. (1996). *Organismus und Ordnung—Zu Genesis und Kritik der Systemtheorie Ludwig von Bertalanffys*, unpublished doctoral thesis, University of Vienna.

Hofer, V. (2000). "Der Beginn der biologischen Systemtheorie im Kontext der Wiener Moderne. Diskurslinien und Wissenschaftsgemeinschaften als intellektueller Hintergrund für Ludwig von Bertalanffy," in K. Edlinger, W. Feigl and G. Fleck (ed.), *Systemtheoretische Perspektiven: der Organismus als Ganzheit in der Sicht von Biologie, Medizin und Psychologie*, Frankfurt/Main, Lang, pp. 137-157.

Hofkirchner W. (2005). "Ludwig von Bertalanffy: Forerunner of evolutionary systems theory," in J. Gu and G. Chroust (eds.), *The New Role of Systems Sciences For a Knowledge-based Society: Proceedings of the First World Congress of the International Federation for Systems Research*, Kobe, Japan, CD-ROM.

Jonas, H. (1951). "Comments on general system theory," *Human Biology*, 23.

Johnston, W.M. "Von Bertalanffy's place in Austrian thought," in W. Gray and N.D. Rizzo (ed.), *Unity through Diversity—A Festschrift for Ludwig von Bertalanffy*, vol. 1, New York, Gordon & Breach.

Kamarýt, J. (1961). "Die Bedeutung der Theorie des offenen Systems in der gegenwärtigen Biologie (Zur Kritik der Philosophie des Organischen bei Bertalanffy)," *Deutsche Zeitschrift für Philosophie*, 9.

Kamarýt, J. (1963). "Ludwig von Bertalanffy a syntetickè smèry v zàpadnì biologii," in J. Kamarýt (ed.), *Filosofickè Problèmy Moderni Biologie*, Praga, Československà Akademie.

Kamaryt, J. (1973). "From science to metascience and philosophy," in W. Gray and N.D. Rizzo (ed.), *Unity through Diversity—A Festschrift for Ludwig von Bertalanffy*, vol. 1, New York, Gordon & Breach.

Kanaev, I.I. (1973). "Some aspects of the history of the problem of the morphological type from Darwin to the present," in W. Gray and N.D. Rizzo (ed.), *Unity through Diversity—A Festschrift for Ludwig von Bertalanffy*, vol. 1, New York, Gordon & Breach.

Kornwachs, K. (2004). "System ontology and descriptionism—Bertalanffy's view and new developments," *TripleC*, 2.1: 47-63.

Laszlo, E. (ed.) (1972). *The relevance of general systems theory—Papers presented to Ludwig von Bertalanffy on his seventieth birthday*, New York, G. Braziller.

Laszlo, E. (1973). "LvB and Levi-Strauss: systems and structures in biology and social anthropology," in W. Gray and N.D. Rizzo (ed.), *Unity through Diversity—A Festschrift for Ludwig von Bertalanffy*, vol. 1, New York, Gordon & Breach.

Lektorsky, V.A. and Sadovsky, V.N. (1960). "On principles of systems research (related to L. Bertalanffy's general system theory)," *General Systems*.

Le Moigne, J.L. (1977). *La théorie du système général—Théorie de la modélisation*, Paris, P.U.F.

Lilienfeld, R. (1978). *The rise of systems theory*, New York, Wiley.

Mulej, M. *et al.* (2004). "How to restore Bertalanffian systems thinking?," *Kybernetes*, 33.1: 48-61.

Müller, K. (1996). *Allgemeine Systemtheorie—Geschichte, Methodologie und sozialwissenschaftliche Heuristik eines Wissenschaftsprogramms*, Opladen, Westdeutscher Verlag.

Needham, J. (1932). "Thoughts on the problem of biological organization," *Scientia*, 26.

Needham, J. (1933). "*Theoretische Biologie*" (review), *Nature*, 132.

Nierhaus, G. (1981). "Ludwig von Bertalanffy 1901-1972," *Sudhoffs Archiv*, 65.

Appendix B: Secondary Sources...

Peter, K. (1973). "Sorokin and von Bertalanffy: a convergence of views," in W. Gray and N.D. Rizzo (ed.), *Unity through Diversity—A Festschrift for Ludwig von Bertalanffy*, vol. 1, New York, Gordon & Breach.

Phillips, D.C. (1976). *Holistic thought in social science*, Stanford, MacMillan.

Pogliano, C. (1987). "Ludwig von Bertalanffy," *Belfagor*, 42.1.

Pouvreau, D. (2005). *Vers une histoire de la 'théorie générale des systèmes' de Ludwig von Bertalanffy*, unpublished D.E.A. thesis, EHESS, Paris.

Pouvreau, D. (2005). "Eléments d'histoire d'une fécondation mutuelle entre holisme et biologie mathématique," *Sciences et techniques en perspective*, 2nd series, 9.2.

Pouvreau, D. and Drack, M. (2007). "On the history of Ludwig von Bertalanffy's 'General Systemology', and on its relationship to Cybernetics—Part I: Elements on the origins and genesis of Ludwig von Bertalanffy's 'General Systemology'," *International Journal of General Systems*, 36.3.

Regelmann, J.P. (1986). "Systemtheorie und Krise," in J.P. Regelmann and E. Schramm (ed.), *Wissenschaft der Wendezeit—Systemtheorie als Alternative?*, Frankfurt/Main, Fischer.

Rizzo, N.D. (1972). "The significance of von Bertalanffy for psychology," in E. Laszlo (ed.), *The relevance of general systems theory*, New York, Braziller.

Rosen, R. (1969). "Putting a science back on the track—*General System Theory*" (review), *Science*, 164.

Ross, D.M. (1973). "LvB—Leading theoretical biologist of the 20th century," in W. Gray and N.D. Rizzo (ed.), *Unity through Diversity—A Festschrift for Ludwig von Bertalanffy*, vol. 1, New York, Gordon & Breach.

Rothschuh, K.E. (1936). *Theoretische Biologie und Medizin*, Berlin.

Russel, E.S. (1933). "*Theoretische Biologie*" (review), *Science Progress*.

Sadovsky, V.N. (1965). "Aspects méthodologiques d'une théorie générale des systèmes," *Studium Generale*, 18.

Saint-Germain, M. (1979). *Étude synoptique de la théorie organismique, de la théorie des systèmes ouverts, de la théorie générale des systèmes de L. von Bertalanffy*, unpublished doctoral thesis, University of Ottawa.

Saint-Germain, M. (1981). "Von Bertalanffy's organismic theory, open system theory, general system theory as an organized system," *General Systems*, 26.

Smith, R.A. (1970). "Review of LvB's *General system theory*," *General Systems*, 15.

Strauss, D.F.M. (2002). "The scope and limitations of Von Bertalanffy's systems theory", *South African Journal of Philosophy*, 21.3.

Taux, E. (1986). "Die Verwendung erkenntniskritischer Begriffe in der theoretischen Biologie Uexkülls und Bertalanffys," in J.P. Regelmann and E. Schramm (ed.), *Wissenschaft der Wendezeit—Systemtheorie als Alternative?*, Frankfurt/Main, Fischer.

Thumb, N. (1943). "Die Stellung der Psychologie zur Biologie: Gedanken zu L.v. Bertalanffys Theoretischer Biologie," *Zentralblatt für Psychotherapie*, 15.

Tribiño, S.E.M. Gorleri de (1945-1946). "Una nueava orientación de la filosofía biológica: El organicismo de Luis Bertalanffy," *Revista del Colegio Libre De Estudios Superiores*, 28.165-166.

Ungerer, E. (1973). "The position of Ludwig von Bertalanffy in contemporary thought," in W. Gray and N.D. Rizzo (ed.), *Unity through Diversity—A Festschrift for Ludwig von Bertalanffy*, vol. 1, New York, Gordon & Breach.

Weckowitz, T.E. (1987). "Ludwig von Bertalanffy's contributions to theoretical psychology," in W. Baker, L.P. Mos, H.V. Roppard and H.J. Stam (ed.), *Recent Trends in Theoretical Psychology*, New York, Springer.

Appendix B: Secondary Sources...

Withers, R.F.J. (1952-1953). "Problems of life" (review), *British Journal of Philosophy of Science*, 3.

Young, J.Z. (1933). "*Modern Theories of Development*" (review), *Science Progress*.

Zeeuw, G. de (2004). "A forgotten message? Von Bertalanffy's puzzle," *Kybernetes*, 35.3-4.

Zerbst, E. (1972). "The impact of von Bertalanffy on physiology," in E. Laszlo (ed.), *The relevance of general systems theory*, New York, Braziller.

Index

A

AAAS (American Association for the Advancement of Sciences) 138-9
Alpbach 82, 84, 90-1, 179
ambiguity 59-60, 79
American Association for the Advancement of Sciences (AAAS) 138-9
American Cancer Society 146-7, 160, 162
applications 27, 37, 46, 55, 63, 66, 71, 82, 97, 122, 131, 149-50, 179
art history 16-17, 23-4, 186
Austria 14, 26, 37, 51, 53, 60, 79, 81, 92, 148, 156-7, 173, 175-6, 179

B

behavior 20, 79, 123, 139, 173-4
  animal 20, 134
Behavioral Sciences 122, 124, 134-6, 151-2, 155, 177
behaviorism 104, 136, 173
Berlin 27-8, 42, 71, 77, 148-50, 154
biological systems 32, 41, 48
biologism 59, 135-6
biology
  mathematical 40, 46-7, 53, 107
  philosophical history of 72, 111-12
  philosophy of 19, 23, 26, 29-30, 37, 95
biophysics 46, 54, 100-1, 120, 148, 156, 163, 186
botany 16, 62, 81, 99
Buffalo 6, 180-1

C

California 51, 114, 121, 124, 128
Canada 11, 88, 94, 97-9, 101, 103, 105, 107, 109-11, 113, 115, 117, 119, 127, 166-7, 175-6
Cancer 107-8, 147, 160-2, 164
CASBS (Center for Advanced Study in the Behavioral Sciences) 124-7, 130-1, 134-5, 137, 139-41, 157-8, 182
cell 32, 44, 108
Chicago 46-53, 114, 121-6
community 59, 137
complexity 2, 4, 6, 32, 43, 136
components 44, 64, 110
conceptions 30, 36, 39, 42, 45, 47, 66-7, 74, 83, 104, 131, 173
  atomistic 57, 59, 68
conflicts 28, 47-8, 50, 69, 77, 83, 118-19, 152, 154, 162
context 27, 55, 64, 69, 74, 112, 124, 135
creation 37, 77, 99, 104, 111, 120, 124, 137, 140, 149, 157, 164, 167
creativity 145, 173-4
critique 31, 56-7, 59, 65, 132, 134-5, 146, 174, 182, 186
culture 17, 21, 25, 59, 68, 83, 85, 117, 130, 133, 144, 170, 181, 186
cybernetics 1, 110, 144, 156, 173-4

D

Darwinism 24, 31, 56-7
Denazification 76-7, 81, 85
denouncing 59, 120
dichotomy 59, 110
differentiation 33, 45, 144-5, 173
disciplines 2, 10, 50, 73, 116, 122, 171, 173

Index

discovery 7, 11, 40, 44-5, 178, 182
diversity 4, 24, 184, 186
dynamic systems 41, 48, 67

E

Edmonton 166-9, 172, 175, 180, 183
education 1, 21, 31, 61, 79, 84-5, 93-4, 111, 124, 140, 145, 170
environment 20, 45, 64, 83-4, 103, 133, 145, 166, 173
epistemological root of scientism 160-1
equilibrium 33, 65, 73, 105, 145, 173-4
  dynamical 44, 53
  homeostatic 110, 148, 173-4
eugenics 31, 56
Europe 96, 125, 141, 147, 149, 152-5, 157, 163-5, 175-7, 179
evolution 7, 17, 19, 30, 41-5, 48, 53, 67, 83, 100, 154, 174
examination 4, 8, 76, 78-9, 87, 146
extension 34, 52, 67, 83, 90, 93, 119, 127, 132, 170, 172, 180

F

feedback 110, 145, 156
Ford Foundation 77, 123-4, 157
framework 8, 32-3, 95, 118, 138, 141-2, 148, 153
functions 33, 44, 46, 98, 110, 137

G

General System Theory 1-2, 11, 184
General Systemology 11, 47, 49, 82, 85, 95, 108, 121, 137, 172, 182
General Systems 130-1, 137

General Systems Theory 137, 139
genes 43-4, 72, 74
genetics 33, 43, 53, 65, 67, 95, 154
Germany 5, 20, 26, 37, 39, 70, 79, 92, 118, 124, 148-50, 153-4, 156-7, 163-4, 172, 175
Great Britain 36, 93, 95
groups 44, 49, 82, 102, 111, 153
Growth 40, 51, 53, 66-7, 99, 106-7, 123, 144
  global 39-40, 106
  relative 66-7

H

heredity 20, 31, 69-70
hierarchization 44, 104
hierarchy 21, 44
human behavior 83, 123, 135-6
human societies 122, 136-7
humanities 1-2, 50, 137, 164, 175

I

individuals 170, 174
Inductive Metaphysics 21, 24
information 110, 154, 175
Innsbruck 16
integration 9, 33, 111, 117, 123, 159
  interdisciplinary 124, 139-40
interaction 41, 92, 97-8, 133, 145, 171, 173
  dynamic 33, 44, 110
isomorphisms 41, 49-50, 171

K

kinetics 41, 67, 100-1

Index

knowledge 5, 10, 16, 21-2, 32, 39, 42, 68, 92, 103, 111, 124, 130, 132-3, 136, 185-6

L

laws 21, 25, 32-3, 40-1, 49-50, 67, 100, 137-8
legitimization 59-60, 65, 135-6, 187
Logic 16, 33, 45, 50, 83-4, 90, 145, 161
Los Angeles 141-3, 145, 147, 149-51, 153, 155, 157, 166

M

materialism 178
 dialectical 178, 185
mathematics 18, 24, 40, 42, 98, 101, 133
Max Planck Society 157, 164
McGill University 95, 97-8
mechanism 25-6, 32, 35
medicine 18, 31, 36, 45, 69-70, 73, 80, 98-9, 119, 153, 164
Menninger Foundation 159-62
Mental Health Research Institute (MHRI) 151
metabolism 30, 40, 66, 71, 81, 105, 107-8
metaphysics 16, 21, 28, 30, 186
Michigan 123, 151
Montreal 95, 97-8, 107, 126
Montreal and Ottawa 97, 99, 101, 103, 105, 107, 109, 111, 113, 115, 117, 119, 121, 123, 125, 127
morphogenesis 20, 25, 38-9, 41
morphology 16, 39-40, 67
movement, systemic 110, 174-5, 182
Munich 81, 149-50, 156-7, 164
mysticism 17, 23-4, 56, 109, 144

N

National Socialism 5, 10, 55-6, 58-9, 69
natural philosophy 16, 37, 125, 149, 154, 172, 175, 186
Nazi ideology 2, 79
Nazis 52, 58, 78, 81
negotiations 7-8, 77, 165
neo-positivists 28, 42
NSDAP (National Socialist Party) 54-6, 60-1, 63, 78

O

opportunism 56, 58, 61, 79, 187
Organic Growth 39, 51, 53, 64-7, 71, 74, 105, 107, 132, 154
organism 17, 20, 24-5, 30, 32, 39, 45, 60, 64-5, 67, 73, 103, 137, 143, 172-4
Organismic 26, 40-1, 44-5, 47, 59, 72, 74, 90-2, 103
Organismic biology 32, 39, 43, 68-9, 136
organismic perspective 35, 39, 41, 43, 65, 82, 164
organismic philosophy 25, 29-31, 35, 37, 39, 41, 43-5, 53, 59-60, 69, 73, 79, 90, 178
organismic principles 44, 104, 110, 145
organismic programme 40-1, 65
Organismic Psychology 144, 171-3
organization 21, 25, 31-3, 41, 49, 55, 58-9, 102-3, 110, 123, 125, 127, 136-9, 178

P

perspectives 32-3, 37, 39, 41, 69, 84, 111, 133, 135, 160, 174, 186
perspectivism 111, 132-3, 145, 161, 168-9
phenomena 38, 40, 43, 133, 143, 161
  biological 25, 39-40

Index

Philosophical Anthropology 82, 90-1, 134, 154, 160, 168
philosophy 1, 4, 16-18, 21, 23, 25, 29, 36-8, 40, 46, 60, 62, 69, 73, 149, 167-9
　perspectivist 68, 130, 132, 183
philosophy of art history 17, 23-4
philosophy of history 17, 186
physicists 46, 96, 101-2, 151
physics 20, 31-3, 50, 95, 100, 132, 171, 176
physiology 16, 30, 38, 40, 52, 67, 99, 107, 120, 141-2, 146, 184
positivism 51, 111, 160
power 24, 56, 60, 70, 78, 83, 88, 133, 162
principles 20, 25, 30, 32-3, 40-1, 49, 85-6, 90, 100, 130-1, 141, 143-4, 153
Project of General Systemology 95, 121, 130
Psychiatry 105, 110, 130, 142-4, 146, 159-60
Psychologists 27-8, 84, 96, 102, 166
psychology 16, 20-1, 31, 50, 103-5, 110, 130, 132, 141, 144-6, 159, 166-7, 171-3
Psychosis 143

R

relationship 39, 42, 47, 53, 58, 83, 105, 114, 119, 181
RNA 108
robots 172, 178, 185
Rockefeller foundation 46, 51-2, 80, 87

S

SAGST (Society for the Advancement of General Systems Theory) 137-9, 150-1
San Francisco 141-2, 168
schizophrenia 143-4
scholarship 32, 37, 39, 46, 88
Scientism 160-1

SGSR (Society for General Systems Research) 130, 150-1
social sciences 50, 92, 122-4, 132, 141, 181
Society 49, 57, 59, 68, 137-40, 151, 161, 170, 174-5, 181
specificity 83, 134
Stanford 114, 121, 127-8, 130, 139-40
State University of New York (SUNY) 180-1, 183
Switzerland 88, 106, 148
symbolic systems 83, 111
symbolism 134, 168, 175
symbols 83, 90, 134
Systemology 11, 48-50, 65, 85-6, 90, 92, 95, 101, 103-4, 108-11, 121-3, 130-2, 170-2, 175, 178, 182-3
systems movement 4, 6-7
systems philosophy 2, 4, 181

T

Theoretische Biologie 32, 35-7, 43, 73, 75
theory, mathematical 39, 66-7, 102
theory of heredity 31, 69-70
theory of knowledge 16, 21, 42, 103, 160, 186
Theory of Organic Growth 39, 53, 64-5, 71, 74, 107
theory of organization 103, 137
thermodynamics 64, 100-1
Third Reich 54-5, 57, 59, 61, 63, 65-7, 69, 71, 73-5, 77
totalitarian State 84, 135-6

U

United States 4, 6, 9, 42-3, 45-7, 51, 53, 64, 87, 96, 98, 113, 115, 122, 125-6, 180-1
unity 21, 27, 40, 50, 84, 111, 138, 184
University of Ottawa 1, 8, 98-9, 106, 113-15, 118-21,

Index

125-9
University of Vienna 8, 11, 17-18, 20, 26, 37-8, 46,
    51, 58, 61, 63, 70, 88, 93-4, 176

V

Vienna Circle 5, 19, 28, 42-3, 50
vitalism 16, 26, 32, 35, 40

W

War 9-10, 58, 60, 69-70, 72-6, 82, 99, 103, 135, 175
Woods Hole 51, 53

www.ingramcontent.com/pod-product-compliance
Lightning Source LLC
Chambersburg PA
CBHW060947230426
43665CB00015B/2097